INVENTING
FLIGHT

INVENTING FLIGHT

The Wright Brothers &
Their Predecessors

John D. Anderson, Jr.

The Johns Hopkins University Press
Baltimore and London

The Johns Hopkins University Press

2715 North Charles Street

Baltimore, Maryland 21218-4363

www.press.jhu.edu

Library of Congress Cataloging-in-Publication Data

Anderson, John David.

    Inventing flight : the Wright brothers and their predecessors / John D. Anderson, Jr.

      p.   cm.

    Includes bibliographical references and index.

    ISBN 0-8018-6874-2 (hardcover : alk. paper) — ISBN 0-8018-6875-0 (pbk. : alk. paper)

    1. Aeronautics—Research—History—19th century. 2. Wright, Orville, 1871–1948—

Knowledge—Aeronautics. 3. Wright, Wilbur, 1867–1912—Knowledge—Aeronautics.

4. Aeronautical engineers—United States—Biography. I. Title.

    TL516.A53 2004

    629.13'009'034—dc21

2003010631

A catalog record for this book is available from the British Library.

Contents

Preface

Most everyone has flown on an airplane, at least once if not often. At the opening of the twenty-first century, airplanes supply a common feature of modern life. We tend, in fact, to take them for granted. They fly overhead, and we hardly make the time to look up. When we do, we see a flying machine that generally adheres to a standard form—fixed wings, a fuselage, and a tail—a machine powered by some type of engine that may be a jet or a reciprocating engine-propeller combination. This configuration, too, we take for granted.

But of course it was not always so familiar. In the late nineteenth century, no one knew exactly what the airplane should look like. So how did the now-familiar shape of the airplane come about? What technological evolution finally culminated in the successful airplane? The answers to these questions take us to an interesting story, the essence of this book.

The narrative of the development of the airplane has been told many times and in many ways—each replete with excitement, human dynamics, personal defeats, and a few important victories. I, too, relate the excitement, but I focus on the development of the early technology of flight. If asked "Who invented the airplane?" many readers would answer, "The Wright brothers, of course." Yet the Wright brothers did not invent the concept of the airplane; rather, they invented the first practical airplane. They stood on the shoulders of giants, and we must become acquainted with those giants.

I wish to acknowledge in general my colleagues in the Aeronautics Department of the National Air and Space Museum for providing a congenial and learned environment in which to explore all matters pertaining to the history of flight. In particular, many stimulating discussions with Tom Crouch, Von Hardesty, Peter Jakab, and Dominic Pisano have served over the years to expand my intellectual horizons. Particular thanks goes to a member of my own profession of aerospace engineering, Professor Walter Vincenti at Stanford University, whose suggestion started me on this writing project, and to Robert J. Brugger of the Johns Hopkins University Press for inviting me to undertake this book. Special thanks go to Arnold Nayler and Brian Riddle who made available the resources of the Library of the Royal Aeronautical Society, London, and who enthusiastically provided me with important research data. I also wish to acknowledge my wife, Sarah-Allen, who has been very supportive of my professional activities in the history of technology, and in particular the history of aeronautical engineering. Finally, I wish to thank my longtime friend and scientific typist, Sue Cunningham, for typing this manuscript.

INVENTING
FLIGHT

Prologue

The Miracle of Flight

Not within a thousand years will man ever fly.

—Wilbur Wright, 1901, in a fit of despair

SUCCESS FOUR FLIGHTS THURSDAY MORNING ALL AGAINST TWENTY ONE MILE WIND STARTED FROM LEVEL WITH ENGINE POWER ALONE AVERAGE SPEED THROUGH AIR THIRTY ONE MILES LONGEST 57 SECONDS INFORM PRESS HOME CHRISTMAS. ORVELLE WRIGHT

—Orville Wright to his father, December 17, 1903; a telegram, with the original misprints

The scene: Windswept sand dunes on Kill Devil Hills, four miles south of Kitty Hawk, North Carolina. The time: About 10:35 A.M. on Thursday, December 17, 1903. The characters: Orville and Wilbur Wright and five local witnesses. Poised, ready to make history, is a flimsy, odd-looking machine, made from spruce and cloth in the form of two wings, one placed above the other; a crude elevator (a horizontal up/down control piece) mounted on struts in front of the wings; and a double vertical rudder (left/right control device) behind the wings. A twelve horsepower engine stands on the lower wing, slightly right of center. To the left of this engine Orville Wright lies flat on the wing, facing into the brisk wind.

Behind him rotate two ungainly looking airscrews (propellers), driven by two chain-and-pulley arrangements connected to the same engine. The machine begins to move along a sixty-foot launching rail on level ground. Wilbur Wright runs along the right side of the machine, supporting the wingtip so that it will not drag in the sand. Near the end of the starting rail, the machine lifts into the air. John Daniels of the Kill Devil Life Saving Station takes a photograph that preserves for all time one of the momentous events in aviation history.

The machine flies unevenly, rises suddenly to about ten feet, then ducks quickly toward the ground. This erratic flight continues for twelve seconds, ending when the machine falls to the sand, 120 feet from the point where it lifted off. Thus ends a flight that, in Orville Wright's own words, was "the first in the history of the world in which a machine carrying a man had raised itself by its own power into the air in full flight, had sailed forward without reduction of speed, and had finally landed at a point as high as that from which it started."[1] Together, the Wrights realized the dreams of centuries and earned the distinction of becoming the premier aeronautical engineers at the time. (Their first successful flying machine, the Wright Flyer I, now rests in the National Air and Space Museum of the Smithsonian Institution in Washington, D.C., as a memorial.)

Today, virtually every winged aircraft (excluding, that is, "flying wings"[2] and helicopters) incorporates the basic elements of the Wright Flyer—wings to create lift, a form of propulsion to produce forward thrust, and moveable horizontal and vertical surfaces that enable the pilot to control the airplane's rolling, pitching, and yawing motion. Whereas the first flight at Kill Devil Hills lasted only twelve seconds, and the longest of the Wrights' subsequent flights that morning covered 852 feet in fifty-nine seconds, by 1905 the brothers were staying in the air for up to thirty-eight minutes and covering twenty-four miles—or until they literally ran out of gasoline.

The Wright Flyer was high-tech for its day. The Wrights conceived their machine as a system, embodying the mutually supportive elements of aerodynamics, propulsion, structures, and flight dynamics. The performance of a modern racing car ultimately depends on its weakest component, and the same is true of a flying machine. No matter how well-designed in terms of aerodynamics, propulsion, and flight dynamics, a machine with a faulty structural element will

Orville Wright at the controls of the Wright Flyer, December 17, 1903. This photograph, taken at the instant of takeoff from the sand of Kill Devil Hills, near Kitty Hawk, North Carolina, records the most historic moment in aviation—the first successful sustained, controlled, flight of a powered, heavier-than-air, piloted flying machine. With this flight began a century of unparalleled development of the airplane. Courtesy of the National Air and Space Museum.

fail, either not getting off the ground in the first place or falling from the air. The Wright brothers succeeded because they tested and developed each component until it adequately performed its function in their system.

Despite the honor due them for ingenuity, dedication, and persistence, Orville and Wilbur Wright did not invent the airplane entirely by themselves. Their flying machine built on timeless human dreams, benefited from the fifteenth-century concepts of Leonardo da Vinci, and relied on a recent century of efforts to solve the technical problems of powered, heavier-than-air flight. The industrial revolution, first in England and then throughout western Europe, provided both mechanical advancements and increasing confidence in technology. Although the physical principles of powered flight remained unknown at the beginning of the

nineteenth century, by the Wrights' day in Britain, France, and Russia, primitive
vehicles were beginning to hop momentarily off the ground. Optimism abounded.
If Orville and Wilbur had never entered the race, most likely someone else, somewhere, would soon have constructed a successful airplane.

Even so, aeronautical technology took several false starts in the nineteenth century and followed many blind alleys. "Failures, it is said, are more instructive than
successes," Octave Chanute, civil engineer and one of the first Americans to take
an interest in powered flight, noted in 1894; "and thus far in flying machines there
have been nothing but failures."[3]

Chanute's remark betrayed frustration (and a little sarcasm). It also illustrated
the engineer's practical approach, and Chanute could easily have gone on to say
something about the distance men of his kind stood from the academic scientists
of his day, who scarcely lifted a finger to help with the flying machine. In 1866, a
group of inventors in Britain, most of them self-trained amateurs, formed an
Aeronautical Society with the hope of fostering discussion and an exchange of
ideas on flying machines. Though they laid the groundwork for the profession of
aeronautical engineering, they did not profit from exchanges with scientists.
Understanding of fluid mechanics, for example, had advanced considerably in the
nineteenth century—thanks to the work of Louis Navier at the Ecole Polytechnic
in Paris, George Stokes in Cambridge, Herman von Helmholtz at various German
universities, and Osborne Reynolds at the University of Manchester. Even so, these
gentlemen, university-educated academicians, had not the slightest interest in flying machines. Lord Kelvin, one of Britain's leading scientists—a student of thermodynamics, heat transfer, magnetism, and electricity but also a practically
minded developer of the transatlantic cable—voiced the prevailing disdain for
aeronautics. "I have not the smallest molecule of faith in aerial navigation," he
declared, "other than ballooning." In 1894, when British aeronaut Hiram Maxim
came forth with yet another heavier-than-air flying machine—a gigantic thing
that lifted only two feet off the ground because it was tethered to a rail—it, too,
proved laughable to some. Kelvin dismissed it as "a kind of child's perambulator
with a sunshade magnified eight times."[4]

Meanwhile, despite the large story of ignorance and failure, George Cayley in
England, Alphonse Penaud and Clement Ader in France, and Otto Lilienthal in

Germany epitomized the trial-and-error approach. On the other side of the Atlantic, after more than a few failures, Samuel Pierpont Langley in 1896 successfully flew a steam-powered, unpiloted machine for more than a minute over the Potomac River, thus proving the technical feasibility of heavier-than-air powered flight. Largely lacking in formal academic training, these men used their intuitive understanding of physical laws in designing and testing the machines they hoped would fly. Experimenting, they won advances. What aeronautical technology did the Wrights inherit from such predecessors? Who played major roles in developing it and why? How much was right? How much was wrong?[5] This book will explore the answers to these questions.

Lord Kelvin, incidentally, died on December 17, 1907, four years to the day after the memorable event at Kill Devil Hills—still denying the viability of flying machines.

Imaginings

To me it seems that all sciences are vain and full of errors that are not born of experience, mother of all certainty, and that are not tested by experience.

—Leonardo da Vinci, *Trattato della Pittura* (Codex Urbinas, 15th century)

The genius of man may make various inventions, encompassing with various instruments one and the same end; but it will never discover a more beautiful, a more economical, or a more direct one than nature's, since in her inventions nothing is wanting and nothing is superfluous.

—Leonardo da Vinci, 15th Century (From manuscript now in the Royal Library
    at Windsor, England)

Imagine relaxing on a beach. The sky is a light blue, making a sharp contrast at the horizon with the dark blue of the water. Profiled against this sky are white seagulls, seeming to fly effortlessly, alternately flapping their wings for speed and altitude and soaring majestically with outstretched wings. Without knowing it, these seagulls offer a perfect example of one of nature's inventions in which "nothing is wanting and nothing is superfluous," as Leonardo da Vinci wrote in the fifteenth century. Moreover, these seagulls have something you do not—the ability to move at will in the third dimension of the sky. Nature invented you differently, giving you the ability to walk and run only in the seemingly two-dimensional space of the earth's surface. Nevertheless, you might have the urge, indeed the dream, of being able to break the bounds of your two-dimensional world and fly effortlessly through the air along with those seagulls.

SETBACKS

If it were possible to travel back in time—escaping newspapers, radio, television, and so on, as well as all awareness of the automobile, airplane, and modern machinery—you would still observe birds in flight and want to fly. What would you do? Would you immediately conceive of a machine made up of fixed wings, a fuselage, and a tail, such as most airplanes appear today? Would you imagine a flimsy biplane made of cloth and wood, such as the Wright Flyer at the turn of the twentieth century? Most likely you would try to emulate the birds. You would fashion some wings out of wood or feathers, strap them to your arms, climb to the roof of your hut, and jump off, flapping madly. If you were stubborn, you might even try it a second time.

In so doing, you would join a large group of would-be flyers labeled *tower jumpers*. Beginning about 400 B.C., numerous humans attempted flight in this manner. The Benedictine monk, Eilmer (sometimes called Oliver) of Malmesbury, in 1020 constructed some wings and jumped from the roof of Malmesbury Abbey. For his effort he received two broken legs; he blamed his failure on not having had tail surfaces (which function like feathers on a wooden arrow) attached to his feet. In the same century, Saracen of Constantinople fitted a cloak with stiffeners, climbed to the top of a tower, and jumped off, attempting to flap and glide. Not as lucky as the monk, he paid for his curiosity with his life. Tower jumping was not limited by social strata or economic divide. In 1742 the Marquis de Bacqueville attempted to fly across the Seine in Paris with wings fixed to his arms and legs. Jumping from a house by the riverside, he fluttered down onto a washerwoman's barge, breaking both his legs. Tower jumpers contributed nothing to the technology of early flight. Indeed, Giovanni Borelli, a seventeenth-century mathematics professor at the University of Messina and later at Pisa, proved that humans did not have the muscular strength to lift themselves and their apparatus (at least at that time) and fly through the air. He noted that the ratio of muscle power to body weight is much larger for a bird than for a human being.[1] Later tower jumpers should have paid more attention to Borelli.

After jumping from the roof of your hut one or two times, you might get discouraged and give up. Most tower jumpers did. Or, you might try to think of a better way. Perhaps you would conceive of a machine in which you could push or pull levers with your arms or pump with your legs, and its wings would flap up and down and lift into the air. Such flapping-wing machines are called *ornithopters*. This idea fascinated Leonardo da Vinci (1452-1519). More than 500 sketches concerning flight survive from da Vinci's notebooks, including numerous designs for ornithopters that look like mechanical birds. These ornithopters had no redeeming aerodynamic value, and no ornithopter has successfully flown under human power. In the development of the technology of early flight, they represented only setbacks.

Da Vinci did, however, introduce intellectual inquiry into the physical fundamentals that governed flight. An artist, sculptor, mathematician, physicist, engineer, and physician, he epitomized the Renaissance man. In his notes, there are numerous flashes of insight about how a flight vehicle moving through the air generates lift and drag. Some of this insight was misguided, and some was remarkably prescient.[2] He first argued that when a surface struck the air (such as when a bird's wing made a downward movement), the air would be compressed below the surface, and this higher density compressed air would tend to support the surface. In the Codex Trivultianus he wrote: "When the force generates more velocity than the escape of the resisting air, the same air is compressed in the same way as bed feathers when compressed and crushed by a sleeper. And that object by which air was compressed, meeting resistance on it, rebounds in the same way as a ball striking against a wall."[3]

Late, toward the end of his life, da Vinci recorded an observation on the flow field over a lifting object (in that case, a bird), one that qualitatively was closer to identifying the actual source of lift. (The flow field over an object is defined as the variation of velocity, pressure, and all other flow properties throughout the space

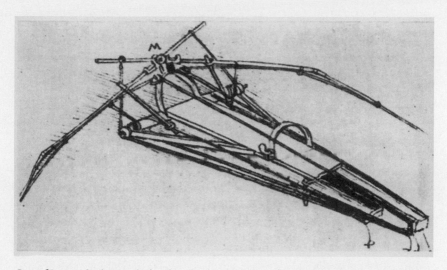

One of Leonardo da Vinci's sketches showing a concept for a human-powered ornithopter, circa 1490. From the *Codex Atlanticus,* housed at the Ambrosiana, Milan.

occupied by the flow.) In Codex E, written around the year 1513, he asked, "What quality of air surrounds birds in flying? The air surrounding birds is above thinner than the usual thinness of the other air, as below it is thicker than the same, and it is thinner behind than above in proportion to the velocity of the bird in its motion forwards, in comparison with the motion of its wings towards the ground; and in the same way the thickness of the air is thicker in front of the bird than below, in proportion to the said thinness of the two said airs."[4] Da Vinci's concepts of "thinner" and "thicker," taken as "lower" and "higher" pressure, contained the earliest inkling that pressure exerted on the top of a moving wing is lower than that exerted on the bottom, hence generating lift. (Lift is that part of the aerodynamic force that acts perpendicular to the forward velocity of the airplane, hence sustaining the airplane in the air.) Da Vinci also believed that the air pressure exerted on the front of the object is higher than that exerted on the back. (Today we refer to this phenomena as pressure drag. Drag is that part of the aerodynamic force that acts parallel but opposite to the forward velocity, hence tending to retard the motion of the airplane. The aerodynamic drag must be overcome by the thrust of the engines to keep the airplane flying.) Da Vinci believed—

reasonably but wrongly—that the aerodynamic force on an object, both lift and drag, depended directly on the speed of the object through the air, that doubling the object's velocity would double its lift and drag.

In the Codex Atlanticus, da Vinci made another observation about air and moving objects. "As it is to move the object against the motionless air, so it is to move the air against the motionless object." In the same set of notes, he phrased the idea slightly differently: "The same force as is made by the thing against air, is made by air against the thing."[5] Da Vinci was the first person to clearly state what later students of flight called the wind tunnel principle, which allows the flow over an airplane in flight through the atmosphere to be studied in a wind tunnel as air blows over a stationary model.

Sophisticated as were da Vinci's thoughts, no one knew of them at the time, and no one experimented with them in practice. His notes, unavailable to others during his lifetime and for long afterward, played no role in the scientific world. Also, his use of reverse, or "mirror-like," handwriting masked the contents of the notes. Da Vinci's work on aerodynamics really came to light only in the nineteenth and twentieth centuries, by which time human flight had advanced well beyond his thinking.

INSIGHTS

Leonardo da Vinci died in 1519. At about that time in Europe, there was a surge of inquiry into the laws that govern the physical world. During this "scientific revolution," Copernicus formulated his heliocentric model of the solar system, which had the planets revolving around the sun (in contrast to the prevailing geocentric theory that the earth was the center of the universe). Gradually, the experimental method of inquiry developed. William Gilbert conducted experiments on magnetism. Francis Bacon's philosophy of learning emphasized the marriage of rational thought with empirical observation—the scientific method. Galileo struggled to piece together a rational theory of mechanics, and René Descartes championed the use of mathematics in the study of physical science. The zenith of this scientific revolution was Isaac Newton's system of rational mechanics that was preserved for the ages in his *Mathematical Principles of Natural Philosophy*, first published in 1687.

Against this background of the scientific revolution, understanding how the aerodynamic force on a body depends on the speed of the body through the air became more important. By prevailing belief, when the velocity doubles, the force doubles; that is, the force is directly proportional to velocity. This seems "logical," although up to the seventeenth century there was no proper experimental evidence or theoretical analysis to say one way or another. Like so much of ancient science, this feeling was based simply on the image of geometric perfection in nature, and what could be more "perfect" than the force doubling when the velocity doubles. Had da Vinci progressed further with his thoughts on flying machines—reaching the point of an actual design on the basis of just the principles known to him—the machine would have had a much larger wing than necessary. Assuming that aerodynamic force (hence lift) is proportional to velocity leads to an inadequate prediction of the force on the wing. The wing must create lift at least equal to the weight of the machine. If the prediction of the lift is too small, and the wing is designed to create a lift equal to the weight, then the resulting wing design will always be too large and weigh too much.

At the end of the seventeenth century, this situation changed dramatically. Between 1673 and 1690, two independent sets of experiments along with Newton's theoretical fundamentals clearly established a new and accurate relationship between speed and force, representing the first major step in the historical evolution of aerodynamics.

One of the chief contributors, the French Edme Mariotte, lived in absolute obscurity until 1666, when he became a charter member of the newly formed Paris Academy of Sciences. Most likely, Mariotte was self-taught in the sciences. He came to the attention of the academy through his pioneering theory that sap circulated through plants in a manner analogous to blood circulating through animals. Within four years, numerous experimental investigators confirmed this controversial theory. Mariotte quickly proved to be an active member and contributor to the academy. His interests were diverse—experimental physics, hydraulics, optics, plant physiology, meteorology, surveying, and general scientific and mathematical methodology. The first in France to develop *experimental* science, Mariotte brought to that country the same interest in experiments that had marked the work of da Vinci and Galileo. Mariotte tried to link existing theory to experiment—a novel thought in that day.

The force produced by various bodies impacting other bodies or surfaces, including fluids, particularly interested Mariotte. He examined and measured the force created by a moving fluid impacting a flat surface. The device he used for these experiments was a beam pivoted at the middle (similar to a seesaw) wherein a stream of water impinged on one end of the beam, and the force exerted by this stream was balanced and measured by a weight on the other end of the beam. The water jet dropped from the bottom of a filled vertical tube, and Mariotte could calculate its speed. In doing so, he drew upon the findings of Evangelista Torricelli, an Italian physicist and mathematician who in 1644 proved that the velocity of a fluid streaming from a hole in the bottom of a tank depended upon the height of the fluid in the tank; that is, the speed equaled the square root of this height multiplied by a known constant. From the results obtained with this apparatus, Mariotte could prove that the force of water on the beam was equal to the flow velocity multiplied by itself (the square of the flow velocity) multiplied by a constant. He presented these results in a paper read to the Paris Academy in 1673. For this work, Edme Mariotte deserves credit for the first major advancement towards the understanding of velocity effects on aerodynamic force.

Christiaan Huygens, another contemporary investigator of speed and force, was born on April 15, 1629, in the Hague, Netherlands, to a prominent family. Tutored by his father until the age of sixteen, Huygens studied law and mathematics at the University of Leiden. Devoting himself to physics and mathematics, Huygens later made improvements in scientific methodology, developed new techniques in optics, and invented the pendulum clock. Huygens, too, served as a charter member of the Paris Academy of Science, moving in 1666 to Paris to more closely participate in its activities. Until 1681, when Huygens moved back to the Hague, he and Mariotte worked, conversed, and argued together as colleagues in the academy. Though recognized as Europe's greatest mathematician, Huygens was reluctant to publish, and he and his work slipped into obscurity during the following century.

In 1668, Huygens began to study the fall of projectiles in a resisting fluid (air or water). Following da Vinci and Galileo, he started with the belief that resistance (drag) was related directly to velocity. Yet one year of analysis and experimental testing convinced him that resistance grew exponentially as speed increased—drag equaled an object's speed times itself—that is—speed squared, times a constant. He made this finding four years before Mariotte published the same result in 1673, but Huygens delayed publishing his data and conclusions until 1690. Eventually, Huygens accused Mariotte of plagiarism, but his jealously only mirrored the nature of collegial work, and he levied this charge after Mariotte's death.

Isaac Newton's theoretical work supported the experimental findings of Mariotte and Huygens later in the same century. Book II of Newton's *Mathematical Principles of Natural Philosophy*, published in 1687, devoted itself to the science of fluid dynamics and hydrostatics.[6] Entitled "The Motions of Bodies (in Resisting Mediums)," it left no doubt of the importance Newton placed on the subject of bodies moving through fluids. He calculated that the resistance acting on a body moving through a fluid acted exactly as Mariotte and Huygens had said.

Newton's interest in fluid dynamics did not derive from any idea of flying machines. It partly grew out of the prevailing practical concerns of naval architecture, in particular the need to predict and understand the hull drag on ships. (By this time the English had clearly demonstrated that a country with a powerful navy could rule large portions of the world. A navy's power depended on the performance of its ships, which in turn partly depended on the ability to understand and predict the drag of ships' hulls.) A much more compelling reason in Newton's mind for calculating the resistance of a body moving through a fluid owed to Descartes's prevailing theory that interplanetary space was filled with matter that moved in vortex-like motion around the planets. Astronomical observations showed that heavenly bodies moved through space in regular and repeatable patterns. If the bodies moved through a space filled with a continuous medium, as Descartes theorized, aerodynamic drag on each body would have to be nothing. Newton meant his fluid mechanics to prove that some measurable drag existed on every body, including the planets as they moved through a continuous medium. This disproved Descartes's theory.

Newton and his contemporaries focused only on the aerodynamic *drag* on an object. They investigated the force on flat plates oriented with their flat surface perpendicular to the flow, or on spheres, or on symmetric bodies at zero incidence angle to the flow (such as the hull of a ship oriented in line with its forward motion). For these cases the net aerodynamic force is indeed all *drag;* such configurations do not generate lift. The wings of flying machines, however, are designed to generate *lift,* and this usually entails some incidence angle (angle of attack) of the wings to the flow. How does the aerodynamic lift vary with regard to the angular orientation of a body in the flow? The works of da Vinci, Galileo, Mariotte, Huygens, or Newton do not address this question. This is somewhat surprising because windmill blades strike the air at an angle of attack, and windmills were in use in Europe since the twelfth century; hence, one might think that angle of attack effects would have been considered for this application. No recorded study, however, either qualitative or quantitative, existed on this question as late as the end of the seventeenth century.

DEVICES

The "scientific revolution" that occurred during the seventeenth century, including the development of Newtonian mechanics, led to knowledge that later would help to reveal the principles underlying powered, heavier-than-air flight. At that time, however, the dream of flying machines in no way spurred any of this inquiry; scientists during the seventeenth and eighteenth centuries had loftier goals in mind—explaining the workings of the universe and exploring the source of the fundamental forces of gravity and magnetism. A few accomplishments during this time, nevertheless, are relevant to our story. These involved *controlled* experiments with specially designed testing equipment in the laboratory. These "ground tests" attempted to measure the aerodynamic force acting on a body as it moves through the air.

One was the invention of the whirling arm by Benjamin Robins in England— a device that was to influence aeronautical testing for the next 150 years. In 1746, Robins succinctly summarized the prevailing understanding of aerodynamic forces in flight in a paper entitled "Resistance of the Air and Experiments Relating to Air Resistance," published in the *Philosophical Transactions* (London): "All the theories of resistance hitherto established are extremely defective, and that it is only by experiments analogous to those here recited, that this important subject can even be completed."[7] Contemporary scientific advances did not immediately lead to the practical calculation of aerodynamic forces on specific body shapes with any

Newton's Velocity-Squared Law

Isaac Newton proved the fundamental relation for aerodynamic force. In the *Mathematical Principles of Natural Philosophy,* he showed that the finite resistance acting on bodies moving through a fluid is "in a ratio compounded of the squared ratio of their velocities, and the squared ratio of their diameters, and the simple ratio of the density of the parts of the system." That is, Newton derived the velocity-squared law, while at the same time showing that the resistance varied with the cross-sectional area of the body (the "squared ratio of their diameters") and the first power of the density (the "simple ratio of the density"). In so doing, Newton presented the first theoretical derivation of the essence of an aerodynamic force. In terms of a modern algebraic equation, Newton's result is embodied in

$$R = c_2 \, \rho \, S \, V^2$$

where R is the force, ρ is the fluid density, S is the cross-sectional area, V is the velocity, and c_2 is a constant of proportionality that depends on the shape and orientation of the body. The value of c_2 has to be measured or calculated for each particular case.

reasonable accuracy. Recognizing the need for some serious experimental advances to fill the void, Benjamin Robins took the lead.

Robins was born to Quaker parents in Bath, England in 1707. He never espoused the Quaker pacifist philosophy; he studied to be a teacher, but gave that up to pursue a scientific career as a military engineer. Intensely interested in mathematics, in 1727 he published a paper in the *Philosophical Transactions* of the Royal Society on a demonstration of the eleventh proposition of Newton's "Treatise on Quadratures." This paper, published in the year of Newton's death, related to the serious controversy between Newton and Baron Gottfried W. von Leibniz over the development of calculus that involved a substantial part of the European scientific community. Robins became a strong supporter of Newton, almost to an extreme; in particular he attacked in writing both Leibniz and several of the Bernoullis who were perceived to be enemies of Newton. In addition to his interest in mathematics, Robins was an experimentalist, and it is in this respect that he made his lasting contributions. Toward the end of his relatively short life, Robins became interested in the investigation of using rockets for the purpose of military signaling.

Robins's contributions to experimental aerodynamics centered around two testing devices that he invented and used for the first time, namely, the whirling arm for measuring aerodynamic forces at low speeds and the ballistic pendulum for examining the aerodynamic characteristics of bodies at high speeds. These devices were used exclusively for testing in air. Robins's interest centered on ballistics—the motion of rockets and artillery shells through the air. He showed no interest in hydrodynamics and ship design, although some of his basic research results applied to hydrodynamics. Robins's whirling arm had an aerodynamic shape mounted at the end of a long horizontal arm. The other end of the arm attached to a vertical shaft that rotated, driven by a falling weight attached to the shaft via a cable-and-pulley system. As the arm rotated, the aerodynamic body moved through the air at some relative velocity and experienced a measured aerodynamic resistance. Whirling arms, however, have a distinct, built-in fallacy; after a period of time of operation, the air in the vicinity of the whirling arm starts to rotate in the same direction as the arm, and it becomes difficult to judge the relative velocity between the moving body and the moving air, hence diminishing the accuracy of any force measurements as a function of relative velocity. Whirling arms provided the only aerodynamic testing devices for the direct measurement of aerodynamic force in the eighteenth century and throughout most of the nineteenth century. Robins also invented the ballistic pendulum, then equally as novel as the whirling arm. Here a projectile fired into a massive pendulum deflects the pendulum, the extent of which is a measure of the momentum (hence velocity) of the projectile.

Using these two devices, Robins made extensive aerodynamic measurements. He verified Mariotte's seventeenth-century finding that aerodynamic force varied with the square of the relative velocity between the body and the airstream

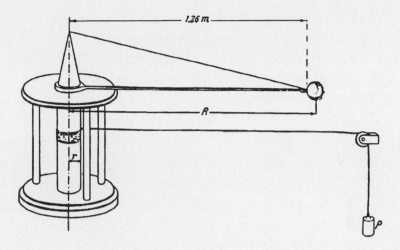

Benjamin Robins's drawing of his whirling arm, 1742. The invention of the whirling arm, a device for generating and measuring aerodynamic force on a body mounted at the end of a rotating arm, is generally credited to Robins. An Englishman working in the field of ballistics, Robins used his whirling arm to measure the aerodynamic drag on artillery shells and rockets. He demonstrated its use before the Royal Society several times in 1746 and reported experimental results obtained with the device in two books published in 1742 and 1746. The arm had a radius of 1.26 meters. A falling weight attached to the shaft of the arm via a cable-and-pulley system rotated the arm. Author's collection.

(at speeds less than the speed of sound). He demonstrated for the first time that two aerodynamic bodies with different shapes but the same projected frontal area have different values of drag; that is, the shape of the body influences the drag. In Robins's time, the prevailing intuition was that drag was mainly due to the projected frontal area of a body and that the shape was secondary. In particular, Robins tested pyramid shapes, first with the apex forward toward the incoming flow, and then with the flat base forward. The latter case produced more drag. He also tested oblong flat plates at 45 degree angles of attack, first with the long side and then with the short side as the leading edge. He found that the drag differed substantially, with the latter case producing more drag. In this respect, Robins discovered for the first time the influence of wing aspect ratio on the aerodynamic drag. (The aspect ratio for a rectangular wing is defined as the wing span, the distance from one tip of the wing to the other, divided by the chord length, the distance from the front of the wing to the back. High-aspect-ratio wings are long and narrow, and small-aspect-ratio wings are short and stubby.) Robins was also the first to observe and note the large increase in drag associated with speeds near the speed of sound. Indeed, when the projectiles were moving at near the speed of sound, from his ballistic pendulum measurements, Robins observed that the aerodynamic force began to vary as the velocity *cubed,* not squared as in the lower-speed cases. Robins had no idea why; he had only his empirical observations of the force. Robins described his aerodynamic results in only two publications, the first a book published in London in 1742, and the second a paper given to the Royal Society in 1746.[8]

Benjamin Robins's work bridged the divide between academic scientists and craftsmen practitioners. Many researchers, including one of the greatest eighteenth-century mathematicians and scientists, Leonhard Euler, read his work. Indeed, Robins's book so excited Euler that he personally translated it into German in 1745, adding some commentary of his own. In 1751, the year of Robins's death, the book appeared in French. Euler extolled Robins's discovery of the large drag rise near the speed of sound and agreed completely with the result. Indeed, Euler and Robins suggested the use of bodies with reduced frontal area at the nose of the body to reduce this drag rise—a purely intuitive idea that proved correct two hundred years later. In a show of appreciation by his contemporaries in England as well as on the continent, in 1747 the Royal Society, Britain's most prestigious, venerable, and entrenched scientific society, whose membership included the best of England's scientists, awarded Robins the Copley Medal.

Eight years after Robins's death, another Englishman, John Smeaton, made a singularly influential contribution to experimental aerodynamics. A professional civil engineer recognized as the first person to make engineering a respected endeavor in the eyes of British society, Smeaton began his education in law. Quickly finding, however, that his talent lay in mechanical matters, he became a successful maker of scientific instruments. The beginning of the industrial revolution in England created the demand for massive civil engineering projects.

Transonic Drag Rise

Mach number is defined as velocity divided by the speed of sound. An object moving through the air at velocity V has a Mach number M defined as M = V/a, where *a* is the speed of sound. The range of Mach numbers near one, typically Mach numbers from 0.8 to 1.2, is defined as the transonic regime. Any object attempting to fly through Mach one experiences a large increase in drag—the transonic drag rise. This increase in drag is caused by shock waves that occur in the flow field around a body moving at close to the speed of sound. The transonic drag rise has a major impact on modern airplanes designed to fly near, at, and beyond the speed of sound. That Benjamin Robins observed the transonic drag rise in the eighteenth century is amazing; the practical use of this knowledge did not come until the middle of the twentieth century. For the discovery of the transonic drag rise, Robins deserves a high place in the history of aerodynamics.

Smeaton took advantage of these opportunities; he designed and constructed several harbors in England and established a reputation as a structural engineer. His rebuilding of the Eddystone lighthouse after two previous contractors failed earned him general fame in England. The Royal Society made him a fellow, and the Society of Civil Engineers, the first professional engineering association, after his death became known as the Smeatonian Society.

Smeaton received the Royal Society's Copley Medal in 1759 for his work on windmills, which are important to our story. With over 10,000 windmills in England at that time, as well as a number of mills driven by water power, Smeaton carried out experiments on the force of air and water on the vanes of such devices. Smeaton adopted Robins's invention of the whirling arm. The arm moved the blades as a set, and the blades themselves rotated individually, thus simulating the actual operation of a windmill in the face of a wind. A cable-and-pulley mechanism activated by a falling weight spun the windmill blades at the end of the whirling arm. The

Royal Society published Smeaton's experimental results in London in 1759.[9] In this paper Smeaton included a table of aerodynamic force measurements on a flat surface perpendicular to the flow. He found that the force was equal to the area of the surface times the square of the velocity times a constant that became known as Smeaton's coefficient. With the surface area expressed in square feet, the wind velocity in miles per hour, and the force in pounds, the numerical value of Smeaton's coefficient was 0.005. This number is important because many aerodynamic experimenters through the nineteenth century used it.

As an historical sidelight, we note that Smeaton did not compile the actual table of aerodynamic forces and velocities that appeared in his 1759 paper. Rather, a Mr. Rouse sent the table to Smeaton. Rouse independently developed and experimented with the whirling arm, the second person after Robins to use such a device. Smeaton clearly referenced Rouse's input. In a small historical injustice, however, throughout the next two centuries subsequent users of these tables always referred to them as "Smeaton's Tables."

Smeaton left behind a major but controversial contribution—information and an empirical relation on the aerodynamic forces on a surface oriented perpendicular to an airstream. Later investigators used and modified these data (principally Smeaton's coefficient) for the estimation of lift and drag on wings and airfoil surfaces. Uncertainties subsequently found in the value of Smeaton's coefficient generated much controversy. During the nineteenth century, nevertheless, many investigators continued to use the number 0.005 for Smeaton's coefficient.

The work of Newton, Robins, and others discussed here provided fundamental knowledge. Prior to the nineteenth century, however, virtually no progress was made on the proper *mechanical design* of flying machines. Nevertheless, human efforts to fly literally got off the ground on November 21, 1783. A balloon, inflated and buoyed by hot air from an open fire burning in a large wicker basket underneath, ascended into the air and carried Pilatre de Rozier and the Marquis d'Arlandes five miles across Paris. The Montgolfier brothers, Joseph and Étienne, designed and constructed the balloon. In 1782, Joseph Montgolfier, gazing into his fireplace, had conceived the idea of using the "lifting power" of hot air rising from a flame to lift a person from the surface of the earth. The brothers instantly set to work, experimenting with bags made of paper and linen, in which hot air from a fire was trapped.

Joseph and Étienne had been carrying on their father's paper manufacturing business in Annonay, so they were familiar with the material and had the skills to construct the balloon. After several public demonstrations of flight without human passengers, including the eight-minute voyage of a balloon carrying a cage containing a sheep, a rooster, and a duck, the Montgolfiers took the big step. At 1:54 P.M., on November 21, 1783, the first flight with human passengers rose majestically into the air and lasted for twenty-five minutes. Soon afterward, in December, the noted French physicist J. A. C. Charles built and flew a hydrogen-filled balloon from the Tuileries Gardens in Paris.

So people finally lifted off the ground! Balloons, or *aerostatic machines* as the Montgolfiers called them, made no technical contributions to heavier-than-air flying machines. They served a major purpose, however, in triggering the public's interest in flight through the air. People could really leave the ground for an extended period of time and sample the environs heretofore exclusively reserved for birds. Balloons provided the only means of flight for more than another hundred years.

Configurations

The whole problem is confined within these limits, viz. To make a surface support a given weight by the application of power to the resistance of air."

—Sir George Cayley, from "On Aerial Navigation," which appeared in *Nicholson's Journal*, November 1809.

Imagine a reasonably educated person living at the end of the eighteenth century who is interested in developing a flying machine. This person typically would not have gone to Oxford or Cambridge, or any university for that matter, but nevertheless he read extensively. He knew something of the theories and findings of the "scientific revolution," especially the rational mechanics of Isaac Newton. He lived under the influence of the industrial revolution that appeared in Britain about 1760 and later swept through Europe, driven in part by the invention and development of the steam engine by Thomas Savery, Thomas Newcomen, and James Watt—all Englishmen. Although in 1800 these engines were heavy, cumbersome, terribly inefficient, and used mainly for pumping water out of deep mines, rapid technical improvements provided a means of power and transportation that justifies describing the nineteenth century as the

Age of Steam. Key developments occurred in textile machines for spinning thread. Construction of the great system of canals cutting across Britain began. In such heady times, a few optimists believed that heavier-than-air flying machines might be technically possible.

Unfortunately, the design of a successful flying machine posed serious problems. Even if a person had an understanding of some of the scientific fundamentals underlying powered flight, there was a complete void in practical engineering approaches to it. The popular image of a flying machine involved flapping wings for lift and thrust—the ornithopter concept exemplified by da Vinci's thinking. Human beings had broken the bond of gravity and lifted into the air in balloons, but balloons contributed nothing to the technical advancement of heavier-than-air flight. A person interested in building a flying machine at the end of the eighteenth century had precious little to work with.

Making things worse, a result logically derived from Newton's *Mathematical Principles of Natural Philosophy* indicated that heavier-than-air flying machines were technically not feasible. Called the "Newtonian sine-squared law," it predicted that the force on a flat surface (such as a flat wing) oriented at an angle of attack to the incoming airflow varied with the square of the trigonometric sine of the angle of attack. At the small angles of attack suitable to flight (believed to be about ten degrees or less), the Newtonian sine-squared law predicted very low values of aerodynamic lift on the wing, leading to the specter of enormously large (and very heavy) wings in order to generate enough lift to balance the weight of the flying machine. The lift could be increased by increasing the angle of attack, which in turn increased the drag. Overcoming the higher drag dictated a larger and heavier engine, requiring an even larger wing to generate higher lift to balance the increase in weight—a mean circle that led to much pessimism about the future of heavier-than-air flight.

Some experimental evidence, however, cast doubt on the accuracy of the Newtonian sine-squared law. For example, in 1777 the French sponsored a series of experimental measurements of the drag on ships' hulls moving in canals. Jean Le Rond d'Alembert, a leading scientist and mathematician, compared these measurements with predictions using the sine-squared law. He found that the law

Newtonian Sine-Squared Law

In Book II of his *Mathematical Principles of Natural Philosophy*, Isaac Newton modeled a fluid flow as a stream of solid particles that impact directly on the surface of a body immersed in the fluid. From this model, he derived the sine-squared law for the force F exerted by the fluid on a segment of the body surface as

$$F = \rho \, V^2 \, S \, (\sin \theta)^2$$

where ρ is the fluid density (mass per unit volume), V is the flow velocity, S is the surface area of the body segment, θ is the angle between a tangent to the surface and the flow direction, and $\sin\theta$ is the trigonometric sine of θ. Applying the sine-squared law to a flat plate inclined to the flow at an angle of attack α, Leonhard Euler in 1745 obtained the equation for the resultant aerodynamic force R acting perpendicular to the plate as

$$R = \rho \, V^2 \, S \, (\sin \alpha)^2.$$

At small angles of attack, the sine-squared law predicts very low values of the aerodynamic force because the sine of a small angle is a small number. For example, for an angle of attack of three degrees, $\sin 3° = 0.0523$, and when multiplied by itself $(\sin 3°)^2 = 0.00274$, a very small value. The small lift on the wings of a flying machine predicted from the sine-squared law discouraged many inventors during the nineteenth century.

gave reasonably accurate results for inclination angles of fifty to ninety degrees, but it was inaccurate for smaller angles.[1] Even Newton in his *Principles* questioned the validity of his model of fluid flow—the model that led to the sine-squared law. At the turn of the eighteenth century, some flying machine enthusiasts did not believe the pessimistic predictions based on the sine-squared law and pressed on. Sir George Cayley in England was one of those.

GEORGE CAYLEY AND THE CONCEPT OF FLIGHT

In a stroke of genius, George Cayley conceived the configuration that would evolve into the airplane. The source of this genius is difficult to explain. He was born into a well-known Scottish family descended from Robert Bruce on his mother's side and from the Normans who invaded England in 1066 on his father's side. His father, Sir Thomas Cayley, gave George much freedom at home and provided tutors for his education. George acquired an early interest in mechanical devices, and he paid the village watchmaker frequent visits. Only nineteen when his father died, George Cayley inherited the extensive family estate at Brompton, becoming the sixth baronet at Brompton Hall. Cayley had the time, education, and resources to think and to experiment. Yet we do not know when Cayley had the "eureka" moment when he first conceived the idea of a flying machine made up of a *fixed wing*, a fuselage, and a horizontal and vertical tail—his seminal contribution to flight. In this regard, Cayley wrote only that "the first idea I ever had on the subject of mechanical flight was at the Southgate about the year 1792." His first experiments on mechanical flight involved a small model helicopter in 1796, and his first ideas on a fixed-wing machine came a few years later.

In 1804 Cayley designed a hand-launched glider approximately one meter in length. (A full-scale model is on view at the British Science Museum in South Kensington, London.) Today, such a glider may resemble the stuff of child's play, but at the time it represented an astonishing technological breakthrough. For the first time a heavier-than-air machine had a fixed wing, fuselage, and horizontal and vertical tail. Cayley proposed a fixed wing to generate lift, a separate mode of propulsion to overcome the "resistance" (drag) of the machine's motion through the air, and vertical and horizontal tail surfaces for directional and longitudinal static stability. Cayley first illustrated his concept in an unconventional manner; in 1799 he engraved on a silver disk an outline of a fixed-wing aircraft. On one side of the disk is a sketch of a machine with a fixed wing, fuselage (occupied by a person), horizontal and vertical tails at the rear end of the fuselage, and a pair of "flappers" for propulsion. The flip side of the disk shows, for the first time in history, a lift and drag force diagram for a lifting wing surface. No larger than an American quarter, this silver disk is now in the collection of the British Science Museum.

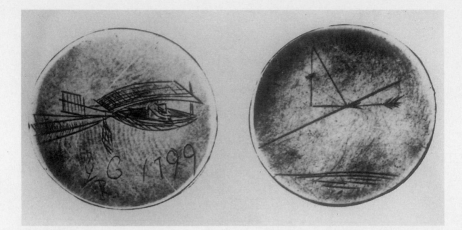

George Cayley's engraved silver disk, the first illustration of the configuration that evolved into the modern airplane, 1799. On one side of the disk, shown at the left, we see a flying machine with a fixed wing, fuselage (occupied by a person), horizontal and vertical tails at the rear end of the fuselage, and a pair of "flappers" for propulsion. On the other side of the disk, shown at the right, is a diagram of the aerodynamic force on a wing (a flat plate) at an angle of attack to the incoming airstream. The arrow shows the flow direction from right to left, and the long diagonal line represents a cross-section of the flat wing. Considering the right triangle above the wing, the hypotenuse represents the resultant aerodynamic force, and the vertical and horizontal sides are the lift and drag, respectively. Courtesy of the National Air and Space Museum.

Cayley's 1804 glider, essentially a kite fixed to a rod-like fuselage, tested his fixed-wing concept. A seminal contribution to the technology of early flight, Cayley described his glider: "A common paper kite containing 154 sqr. inches was fastened to a rod of wood at the hinder end and supported from the fore part from the same rod by a peg, so as to make an angle of 6°. With it this rod proceeded on behind the kite and supported a tail, made of two planes crossing each other at right angles, containing 20 inches each. This tail could be set to any angle with the stick. The centre of gravity was varied by sticking a weight with a sharp point into the stick." In regard to the glider's flight, Cayley wrote: "It would skim for 20 or 30 yards supporting its weight, and if pointed downward in an angle of about 18°, it would proceed uniformly in a right line for ever with a velocity of

Cayley's drawing of his 1804 hand-launched glider, the first fixed-wing configuration to fly. Essentially a common paper kite, it had a wing of 154 square inches. The tail consisted of vertical and horizontal surfaces that could be set at pre-determined angles relative to the rod. A small weight fixed below the nose adjusted the center of gravity to stabilize the glider. Author's collection.

15 feet per second. It was very pretty to see it sail down a steep hill, and it gave the idea that a larger instrument would be a better and a safer conveyance down the Alps than ever the surefooted mule, let him mediate his track ever so intensely. The least inclination of the tail towards the right or left made it shape its course like a ship by the rudder."[2]

Cayley designed his 1804 glider using aerodynamic data for lift and drag obtained from his whirling arm tests. Benjamin Robins invented the whirling arm, and John Smeaton used such a device to test model windmill sails, but Cayley for the first time used a whirling arm for aeronautical purposes.

Cayley carried out his aeronautical work while he was a moderately well-to-do Yorkshire country squire. He did not have a formal college education, but he was exceptionally well-read. This, combined with his enthusiasm for knowledge and invention, made Cayley sought after as one of England's leading scholars in matters of science, technology, and social ethics by the early 1800s.

His seminal contributions to aeronautics formed only part of a wider range of technical and social accomplishments. In 1825 he invented a land vehicle moving on caterpillar treads, the forerunner of the Caterpillar tractor and twentieth-century military tanks. In 1847 he created an artificial hand, a breakthrough in such devices, replacing the simple hook that had been in use for centuries. Working with purely humane purposes, he expected and received virtually no financial compensation for this invention. In 1832 he became a member of Parliament, and in 1839 he founded the Polytechnic Institution in Regent Street, London. In concert with his aeronautical interests, Cayley carried out extensive work on the design of internal combustion engines. He recognized that existing steam engines, with their huge external boilers, were much too heavy in relation to their power output to be of any practical application to flying machines. To improve on this situation, Cayley invented the hot-air engine in 1799 and spent the next fifty-eight years trying to perfect the idea, along with a host of other mechanical designers. The invention of the successful gas-fueled engine in France in the mid-1800s finally superseded Cayley's hot-air engine.

For reasons not completely known, Cayley directed his aeronautical interest to lighter-than-air balloons and airships during the period from 1810 to 1843, making contributions to the understanding of such devices and inventing several designs for steerable airships. By that time the use of balloons had become widespread, even being used in war beginning with the French revolution in 1789. From 1843 until his death in 1857, Cayley returned to the airplane, designing and testing several full-scale craft. One machine with triple wings (a triplane) and human-actuated flappers for propulsion made a floating flight off the ground in 1849, carrying a ten-year-old boy for several yards down a hill at Brompton. The flappers reflected Cayley's misguided propulsion idea, building on his original flapper concept shown on the 1799 silver disk. In 1853 his single-wing (monoplane) glider flew across a small valley (no longer than five hundred yards) with Cayley's coachman aboard as an unwilling pilot. At the end of this flight, the coachman pleaded "Please, Sir George, I wish to give notice. I was hired to drive and not to fly."[3]

WILLIAM SAMUEL HENSON AND THE AERIAL STEAM CARRIAGE

In 1835, a thirty-year-old mechanic and lace machinery operator at Chard, in Somerset, England, dreamed of making a flying machine. William Samuel Henson had a talent for the ingenious design of mechanical devices and already held several patents in his name. Beginning in 1840 he experimented with model gliders, and in 1842 he obtained a patent for the design of a large passenger-carrying flying machine. (British patent law in the nineteenth century did not require demonstration that an invention actually worked.) Henson knew of his contemporary George Cayley's work. A beautiful rendering of Cayley's seminal concept of the modern configuration airplane, Henson's flying machine had a fixed wing, fuselage, and tail and was powered by a steam engine driving two propellers mounted behind the wing. The machine personified Cayley's idea of using a wing to produce lift and a separate propulsive mechanism to produce thrust. Although it embodied this idea, Henson's flying machine was much larger and more aesthetically pleasing than Cayley's designs. Widely published prints of Henson's Aerial Steam Carriage branded on the mind of the general public just what a heavier-than-air flying machine should look like—fixed wings, fuselage, tail, and an engine-propeller combination for thrust. Henson's Aerial Steam Carriage, however, was never built and hence never flown.

Henson, with his friend and fellow lace-making engineer, John Stringfellow, attempted to form the Aerial Transit Company with the financial backing of such men as D. E. Colombine, the Regent Street attorney who had negotiated Henson's patent; John Marriott, a journalist whose value was that he "knew a Member of Parliament"; and a Mr. Roebuck, who was expected to promote a bill in Parliament for a shareholders company to operate an Aerial Steam Carriage. While efforts to officially form the company progressed, Henson and Stringfellow busied themselves with developing the machine. Following Cayley's example, they turned to model testing. In 1843, they obtained the help of John Chapman, a mathematician who also had a whirling arm device. Chapman made more than two thousand recorded aerodynamic experiments on the whirling arm for Henson and Stringfellow. Using these data, they built a model with a twenty-foot wingspan and a wing area of 62.9 square feet, powered by a nicely working small steam engine designed primarily by Henson but improved by Stringfellow. From 1845 to 1847 they tested this model at Balsa Down, near Chard, but they could never get the machine to sustain itself after launching down a ramp. The financial supporters got cold feet. Henson became discouraged and gave up. He married and migrated to the United States in 1848.

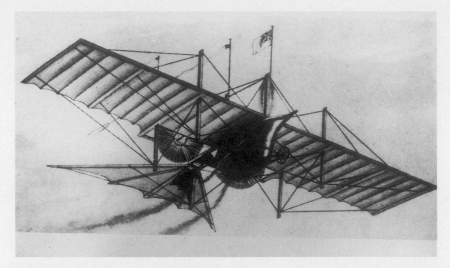

William Samuel Henson's Aerial Steam Carriage. Henson obtained a patent for this grand flying machine in 1842. The machine had a fixed wing, fuselage, tail, and a steam engine driving two propellers behind the wing. The design embodied George Cayley's seminal concept—a fixed wing to produce lift and a separate propulsive mechanism to generate thrust. Although never built or flown, Henson's Aerial Steam Carriage—by means of pictures distributed worldwide—served to permanently brand on the mind of the public the type of configuration that would evolve into the modern airplane. Courtesy of the National Air and Space Museum.

By its design, the Aerial Steam Carriage would have been a large machine with a wingspan of 150 feet and a wing area of 4,500 square feet. (By comparison, the wingspan and area of the popular Boeing 737 twin jet transport are only 95 feet and 1,135 square feet, respectively.) From Cayley's data, Henson reasoned that a square foot of wing area would generate about a half-pound of lift. Including the tail area of 1,500 feet square with the wing area of 4,500 square feet, Henson expected the craft to lift a total of 3,000 pounds. Unfortunately, he lacked lightweight but strong materials for construction of the structure of the machine, and there was no precedent for the construction of lightweight structures to fall back on. He calculated that the engine would have to produce twenty-five to thirty horsepower to propel this heavy flying machine into the air.

Power Loading

Power loading is defined as the ratio of the weight of the airplane to the horse-power of the engine. It is a critical design parameter for an airplane. Everything else being equal, the smaller the power loading (more power and less weight), the faster the airplane will fly, the faster it will climb to higher altitudes, and most importantly, the easier it will take off from the ground (shorter takeoff ground roll). For these reasons, comparing the power loadings of various nineteenth-century flying machines is a valid measure of their capability to get off the ground.

William Samuel Henson's Aerial Steam Carriage had a power loading of one hundred pounds per horsepower. By comparison, the power loading of the Wright flyer in 1903 was 62.5 pounds per horsepower, which was barely small enough to get it into the air. Later, a typical power loading for a World War I biplane such as the Sopwith Camel was about eleven pounds per horsepower. So by comparison Henson's Aerial Steam Carriage was grossly underpowered for its heavy weight, and on this characteristic alone it is extremely unlikely that it would have flown.

A prescient favorable design feature of the Aerial Steam Carriage is the relatively high aspect ratio of its wing. The aspect ratio is a geometric property of a wing, defined as b^2/S, where b is the wingspan (distance from one wing tip to the other) and S is the planform area (the projected area you see when looking directly down on the wing from above). If the wing is rectangular in shape, the aspect ratio is simply b/c, where c is the chord length (distance from the leading edge to the trailing edge of the wing). High-aspect-ratio wings are long and narrow, and small-aspect-ratio wings are short and stubby. At speeds less than the speed of sound, an airplane with a high-aspect-ratio wing will have more lift and less drag than an airplane with a low-aspect-ratio wing, everything else being equal. The beneficial effect of high aspect ratio was not understood in the early part of the nineteenth century. The aspect ratio of the kitelike wing of Cayley's 1804 glider was about one; the aspect ratio of the machine etched on Cayley's silver disk was 1.27. Obviously, George Cayley dealt with almost square-shaped wings. However, Henson's Aerial Steam Carriage had an aspect ratio of five, much higher than any machine seen before. Did Henson have a technical understanding of the beneficial aerodynamic performance of high-aspect-ratio wings? Apparently not. He obtained the patent for the Aerial Steam Carriage in 1842. John Chapman carried out his aerodynamic experiments for Henson and Stringfellow a year later. Even if the experiments had included aspect ratio effects, they would have been too late to affect the design. In any case, there is no evidence that these whirling arm experiments examined the effect of aspect ratio, and nothing existed in the aerodynamic state of the art inherited by Henson that would have allowed him to design for a high aspect ratio. The high aspect ratio of the Aerial Steam Carriage was most likely meant as an aesthetic feature—part of the overall beauty of this design compared to previous concepts for flying machines. Henson had no idea that he was using good aerodynamics. Neither did anybody else.

George Cayley commented on Henson's Aerial Steam Carriage in the April 1843 issue of *Mechanics Magazine.* In terms of the technology of early flight, Cayley made several important points. He criticized the high-aspect-ratio wing of the Aerial Steam Carriage, saying that it would fail structurally under the forces of flight. Given the lack of understanding of aircraft structural design at that time, he was most likely correct. He suggested instead that the total wing area be distributed over two or three smaller wings mounted above each other; that is, he recommended the biplane or triplane configuration. In 1849 Cayley designed a triplane, called Cayley's boy carrier because it made several halting test glides with a young boy on board. In this respect, the design concept of biplanes and triplanes originated with George Cayley. He recognized the need for an airplane to remain stable in roll (rotation about the longitudinal axis that extends from the front to the back of the fuselage). To achieve stability, Cayley recommended for the wings a V form (when seen from the front). This formation is called a dihedral, an essential design feature for stability in roll for many airplanes. Cayley used the term *lateral stability,* a term still used today in association with such rolling motion. Finally, in his criticism of the Aerial Steam Carriage, Cayley used no specific technical data or calculations, instead referring to his own general experience and intuition. Indeed, he could do little else. Here is an excellent example of the still technically undeveloped state of the art of airplane design in the first part of the nineteenth century. Intuition ruled supreme.

Henson's Aerial Steam Carriage was to be powered by two propellers driven by a steam engine. Whereas George Cayley continued to struggle with inappropriate modes of propulsion, Henson had no such problem. He simply came to the most obvious and realistic choice for that time. The steam engine, although still a heavy device, seemed to be the only feasible choice for an aeronautical prime mover.

A review of the development of the steam engine is important at this stage of our story. The steam engine can be traced back to Thomas Savery, who in 1699 made the first use of steam to extract water from mines in England. Savery's device consisted of a large oval vessel filled with steam. When the steam was condensed by pouring cold water on the outside of the vessel, a partial vacuum was formed inside the vessel. This vacuum was transmitted through a long pipe, one end of which was inserted into the pool of water at the bottom of the mine shaft. The pressure of the surrounding atmosphere in the mine shaft pushed the water up the pipe and out of the mine. In 1712, Thomas Newcomen improved on this idea by adopting a cylinder-and-piston arrangement. Steam was introduced into the cylinder underneath the piston and then condensed by a jet of cold water injected into the cylinder, forming a partial vacuum on that side of the piston. Atmospheric pressure was impressed on the top of the piston, driving it down. The piston was connected to one end of a large rocker beam, the other end of which was connected to a pump that extracted water from the mine.

The steam engine came into its own in 1765 when James Watt formulated the idea of a condenser separate from the cylinder. While a laboratory assistant at Glasgow University, Watt recognized that Newcomen's engine wasted an enormous amount of energy because the cylinder was cooled during each stroke of the piston. When the steam was condensed separately, the cylinder remained hot. Watt made a second major improvement, using the expansive power of steam rather than air to push down the top of the piston. Watt knew intuitively that his design increased the thermal efficiency of the steam engine, but the science of thermodynamics was not developed until almost a century later, and no theoretical understanding existed of the basic principles underlying the steam engine. (Watt found by experiment in 1782 that a good "brewery horse" could work at a rate equivalent to raising a weight of 32,400 pounds by one foot each minute. A year later, he standardized the figure of 32,000 foot-pounds per minute as one horsepower, and the firm of Boulton and Watt began classifying their steam engines for sale with ratings of horsepower.) Watt also designed a device for converting the translational motion of the piston into rotary motion for powering rotating machinery.

Richard Trevithick took the next step; in 1802 he used *high-pressure* steam to drive the piston. His steam engines operated at pressures up to 145 pounds per square inch, ten times atmospheric pressure. This not only improved the thermodynamic efficiency, but it allowed engines of a given horsepower to be smaller in size. In 1829, a relatively compact steam engine powered the first successful train locomotive, George and Robert Stephenson's "Rocket." The best measure of improvement in the steam engine is its "duty," a measure of its efficiency then defined as the amount of useful work produced by the engine per bushel of coal. By 1844, the average duty was 68 million foot-pounds per bushel of coal, compared to 6 million foot-pounds for a Newcomen engine in 1767.

So it is no surprise that Henson chose a steam engine for his Aerial Steam Carriage. Steam engines became the aeronautical power plant of choice during most of the nineteenth century. Still somewhat cumbersome and heavy, great effort went into reducing the weight of steam engines for aeronautical use. In addition, Henson's use of an aerial propeller to convert engine power into forward thrust for his flying machine seemed a natural extension of the screw propeller just coming into use on steamships. Steam engines had been used to propel ships with commercial success since 1807, when Robert Fulton's paddle-steamer *Clermont* plied the river between New York and Albany. The steamship experienced a major improvement with the substitution of screw propellers in place of paddles. England built the first successful screw steamer, named the *Archimedes* and weighing 237 tons, in 1838, just four years before Henson obtained the patent for his Aerial Steam Carriage. In regard to the engine and the propellers, Henson's Aerial Steam Carriage exemplified the transfer of technology between different devices.

Stringfellow continued his experiments on models after Henson left England. He exhibited a model steam-powered triplane—reflecting Cayley's influence—at the first aeronautical show in history, an exhibition sponsored by the newly formed Aeronautical Society of Great Britain at the Crystal Palace in London in June 1868. The model hung underneath a small trolley attached by pulley-type wheels to a long cable stretched above the heads of visitors. The best the model could do was to run along this cable. Like Henson's Aerial Steam Carriage, Stringfellow's triplane was mainly influential for its worldwide publicity value. Illustrations of the triplane appeared throughout the end of the nineteenth century.

ALPHONSE PENAUD AND INHERENT STABILITY

A particularly tragic figure in nineteenth-century aeronautics, the engineer Alphonse Penaud glimpsed a promising future in the design of flying machines, one that ended abruptly. Born in Paris in 1850, son of a French admiral, Penaud had a serious hip disease that prevented him from following his father into the navy. Instead, he obtained an excellent engineering education and at an early age devoted himself to aeronautics. Penaud studied the necessary design features that would ensure the inherent stability of a flying machine—features that would tend to restore the machine to equilibrium flight after being disturbed by an eddy or gust in the atmosphere. Cayley also had been concerned with this matter, but Penaud did not know of Cayley's work until much later in his short career.

In 1871 Penaud built a small model aircraft with a pusher propeller driven by twisted strands of rubber that stretched almost the entire length of the sticklike fuselage. The first to use rubber bands for power, Penaud called his model a *planophore*. Weighing only 0.56 ounces with a wing surface of 0.53 square feet and a wing span of eighteen inches, the planophore had a main wing forward and a horizontal tail at the rear. Penaud exhibited the planophore in August 1871 to members of the French Society of Aerial Navigation in the Tuileries garden. The model flew for 131 feet, staying in the air for eleven seconds with 240 turns of the rubber. Of particular importance, Penaud set the horizontal tail at a negative incidence angle of minus eight degrees relative to the chord line of the wing; that is, the tail lay inclined with the front edge downward by eight degrees. This was in

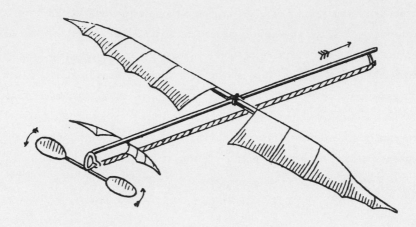

Alphonse Penaud's rubber-band powered airplane, 1871. It had the ability to return to its original equilibrium flight orientation after encountering a disturbance such as a wind gust. Called the *planophore,* this model successfully flew on August 18, 1871, in the Tuileries Gardens in Paris, covering 131 feet in eleven seconds. It became the first inherently stable airplane to fly. Author's collection.

contrast to Cayley's 1804 glider, which had a positive incidence angle relative to the chord of the wing; in other words, Cayley had inclined the tail with the front edge upward. A negative tail setting angle for a rear-mounted tail is necessary for the longitudinal balance of an airplane, and most airplane designs throughout the twentieth century had such a negative tail incidence angle. Penaud is responsible for this design breakthrough.

The first to understand the nature of stability and balance of a flying machine, Penaud located the wing along the fuselage such that its center of pressure (the point through which the lift force on the wing effectively acts) was *behind the center of gravity* of the whole machine. He knew that the lift of the wing must act behind the center of gravity in order to have longitudinal stability. This arrangement would cause the nose of the airplane to pitch down, but the negative tail setting angle results in a *downward* lift on the tail, which tends to pitch the nose up, thus balancing the airplane. When the nose of the airplane is suddenly pitched (rotated) upward by a disturbance such as a gust of wind, the wing lift is momentarily increased due to the increased angle of attack, and the download on the tail is momentarily decreased, both acting to pitch the nose back down, hence restoring the equilibrium. An observer accurately explained this behavior of Penaud's planophore in an account given to the 1874 annual meeting of the Aeronautical Society of Great Britain: "By the alternate action of the weight in front and the rudder [horizontal tail] behind the plane [wing], the equilibrium is maintained. The machine during flight, owing to the above causes, describes a series of ascents and descents after the manner of a sparrow."[4]

The fundamental understanding of the design criteria for longitudinal stability and balance comprises Penaud's major contribution to aeronautics. Cayley's 1804 glider, with its rather large positive tail setting angle necessitated by the extreme forward location of the wing and the center of gravity *behind* the center of pressure of the wing, was a longitudinally unbalanced airplane. Penaud set the standard and provided the fundamental understanding of what was needed for stability and balance. Moreover, Penaud worked out the theory and the practice of stability.

With the assistance of his mechanic Paul Gauchot, Penaud went on to design a large, full-size flying machine, for which he received a patent in 1876. This machine embodied all of Penaud's ideas for the airplane. It was a two-seat monoplane with two tractor propellers (propellers mounted ahead of the wing and oriented to *pull* the airplane through the air, in contrast to pusher propellers mounted behind the wing that *push* the airplane through the air). The propellers rotated in opposite directions to cancel the torque effect of each. (A rotating propeller causes an equal and opposite reaction on the airplane tending to rotate the airplane in the opposite direction, called the torque effect.) Probably for aesthetic appeal, the wings had an elliptical planform. (Penaud had no way of knowing, but an elliptic planform results in optimum aerodynamic efficiency—more lift and less drag—a result discovered in 1918.) An airfoil section of a wing is the cross-sectional shape obtained by cutting the wing with a plane that is parallel to the fuselage and perpendicular to the wing. The airfoil sections of Penaud's wing were cambered (curved, rather than flat). Penaud's reason for choosing this form is unclear; perhaps he copied birds' wings, which are cambered. The wings had a small dihedral angle of two degrees for lateral stability. The machine had two elevators at the rear and a fixed vertical

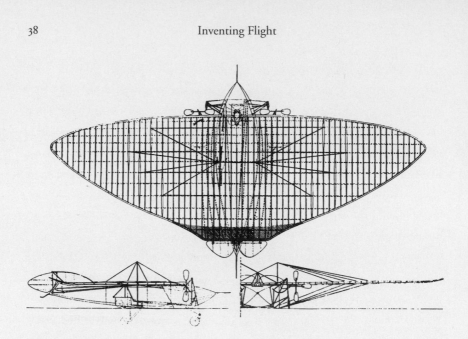

Patent drawing for a piloted, heavier-than-air, powered flying machine designed by Alphonse Penaud with the assistance of his mechanic Paul Gauchot. The design embodied features that would appear in the modern twentieth-century airplane, such as an enclosed cockpit and retractable landing gear with shock absorbers. Although Penaud could not find funding for its construction, the design represents the first full-size airplane with inherent stability. Author's collection.

fin with an attached vertical rudder. The cockpit had a glass dome, a single control column to operate the elevators and rudder, and instruments such as a compass, a level, and a barometer (for measuring altitude). Penaud included retractable landing gear with shock absorbers, and a tail skid. The machine also sported pontoons because Penaud felt that any full-scale flight experiments should be conducted over water. The Penaud-Gauchot machine, including two aviators, weighed an estimated 2,640 pounds (an estimate with little factual basis at that time). Penaud calculated that it would require twenty to thirty horsepower to fly the machine at sixty miles per hour, and that the angle of attack of the wings to the airflow would be about two degrees at that speed. With eighty-eight pounds per horsepower, Penaud's flying machine fell in the same class as Henson's Aerial Steam Carriage.

Penaud never had the chance to fly his machine. He could not find the funds to construct it; he found only criticism and belittlement for his efforts. The general public in the nineteenth century believed that only madmen would try to build a heavier-than-air piloted flying machine. In addition to the tremendous technical challenges of flight, serious investigators and inventors had to deal with this adverse public attitude.

Discouraged, depressed, and in ill health, Penaud committed suicide in October 1880 at the relatively young age of thirty. He nevertheless had made his mark in aeronautics. His work became widely known and continued to influence other inventors into the twentieth century. Unlike Cayley's work, which for unexplainable reasons had became obscure by the 1870s, Penaud's aeronautical investigations were well known by all investigators after him. There is some irony in the fact that, early on, Penaud did not know of Cayley's work and hence repeated some of it but later became responsible for a revival of interest in Cayley and his aeronautical results.

In many respects an aeronautical genius, Penaud, had he lived, might have catalyzed or even achieved successful powered flight before the Wright brothers. We will, of course, never know. Cayley clearly jump-started the development of the airplane at the beginning of the nineteenth century, but it still took another hundred years for the first *practical* airplane to be invented.

Experiments

Now let us consider the nature of the mud in which I have said we are stuck. The cause of our standstill, briefly stated, seems to be this: men do not consider the subject of "aerostation" or "aviation" to be a real science, but bring forward wild, impracticable, unmechanical, and unmathematical schemes, wasting the time of the Society, and causing us to be looked upon as a laughing stock by an incredulous and skeptical public.

—From the fifth annual report of the Aeronautical Society of Great Britain, 1870

Successful flight did not happen until researchers learned enough about its fundamentals that the designers could base their plans on more accurate information. The nineteenth century was the century of aeronautical experimentation, in regard to attempts to achieve actual flight (today we would call this "flight testing") and to ground-based laboratory tests. This chapter tells both stories.

FLIGHT TESTING

Every attempt to get off the ground and achieve sustained flight in a heavier-than-air, powered, piloted machine in the nineteenth century failed. Even so, people learned from each of these tests. These failures represent important stories in the developing technology of early flight.

Felix du Temple designed the first powered, piloted flying machine to get off the ground, but it did not achieve sustained flight. A French naval officer, du Temple received in 1857 a patent for a large flying machine with swept-forward wings covered by silk fabric stretched by several curved spars in the span direction. Du Temple understood the need for lateral stability because his wings had a small degree of dihedral. The wings attached to a short, stubby fuselage containing a motor. The patent did not specify the type of motor. Within a year after his patent came through, du Temple began testing small models of his design, some of which achieved sustained powered model flight with some type of clockwork mechanism engine.

With the aid of his brother, Louis du Temple, Felix eventually constructed a full-size flying machine based on his patent and model tests. This machine, astoundingly, weighed about a ton. Its power plant produced six horsepower, which du Temple had estimated would be sufficient to propel the machine (with a power loading of 333 pounds per horsepower). In 1874 du Temple sent this machine, piloted by a young sailor and powered by some type of hot-air engine (pistons driven by hot air heated by an external energy source—the precise type is unknown), down an inclined ramp at Brest, France. Woefully underpowered, it left the ground for a moment but did not come close to sustained flight, constituting only a powered hop, the first in history. (To qualify as a successful powered flight, a flying machine must sustain itself freely in a horizontal or rising flight path—without loss of airspeed—beyond a point where it could be influenced by any momentum built up before it left the ground. In addition, satisfactory equilibrium and control must be maintained throughout the flight.) Du Temple and his brother continued to work on a lightweight engine without success.

Meanwhile, in Russia, Alexander F. Mozhaiski, a captain of the Imperial Russian Navy, studied the flight of birds and began to experiment with kites. Some were large enough to carry him into the air when a horse-drawn carriage pulled them at high speed. In 1877 he tested a small model with three propellers powered by wound springs similar to a clock. A host of spectators, including members of the Russian Academy of Sciences, observed these short flights on the grounds of the St. Petersburg Riding School. Based on the success of these flights, a committee of scientists from the Academy approved Mozhaiski's design and encouraged the construction of a full-size machine for piloted flight. Patented in 1881 and completed in 1883, this flying machine partially reflected William Henson's design for the Aerial Steam Carriage—a large steam-powered monoplane with a cruciform tail, with one larger tractor propeller in front of the wing and two smaller pusher propellers at the trailing edge of the wing. In 1884, with I. N. Golubev as the pilot, Mozhaiski's flying machine rolled down an inclined ramp at Krasnoye Selo, near St. Petersburg, and became airborne for a distance of about eighty feet. Much later, Soviet historians heralded this test as the first successful airplane flight, but it was actually the second powered hop in history. After 1884, Mozhaiski dropped out of the race for a controllable flying machine.

LABORATORY EXPERIMENTS

In the nineteenth century, scientists and would-be designers of flying machines occupied different worlds. The science of fluid dynamics had made four pivotal advances, about which experimenters with airplanes neither knew nor cared. At the St. Petersburg Academy in the 1730s, Daniel Bernoulli experimented with the relation between velocity and pressure in a fluid, pointing out that in a given flow field, as the fluid velocity increases, its pressure decreases, and vice versa. Although he gallantly tried, Bernoulli never succeeded in precisely quantifying this effect, later called the *Bernoulli principle*. His friend and colleague, Leonhard Euler, also in St. Petersburg, in 1753 used the principles of the newly developed calculus to derive *Bernoulli's equation,*

$$p + {}^1\!/_2 \, \rho \, V^2 = \text{constant},$$

where p denotes pressure, ρ the density of the fluid, and V the velocity. Clearly from this equation, as V increases, p must decrease in order for the sum of the two terms to remain constant. Bernoulli's equation is perhaps the most famous from classical fluid dynamics. Inherent in the Bernoulli principle is the physical explanation of how lift is generated on the wing of a flying machine, but nobody made that connection at the time.

When a fluid (liquid or gas) moves over a body or through a duct, a field of flow establishes where the flow properties (such as pressure, velocity, and density) take on different values at different spatial locations. The quantitative values of these flow-field variables must obey the basic laws of physics, namely, mass conservation (mass can be neither created nor destroyed), Newton's second law (force equals mass times acceleration), and energy conservation. The manner in which these laws are expressed mathematically for a flow is in the form of some elegant and complex partial differential equations called the Navier-Stokes equations. Their solution yields the variation of the flow-field variables as a function of spatial location and time.

These equations have elements that account for internal frictional stresses within the flow. Claude-Louis-Marie-Henri Navier first derived them in 1822 at the Ecole des Ponts et Chaussees in Paris. Navier used a flawed model of the effects of friction, but he came up with the right results anyway. Quite independently, and without knowledge of Navier's work, Sir George Stokes at Cambridge in 1845 properly modeled the effects of friction and obtained the same equations. Thus they are called the *Navier-Stokes* equations. Formulating these equations is one thing; solving them is quite another. To this day, no general analytical solution exists; we have only analytical solutions for a limited number of specialized flows. (In the last quarter of the twentieth century, numerical solutions of the Navier-Stokes equations have been obtained for the flow field over wings and whole airplane shapes, but these require a supercomputer.) Before the mid-nineteenth century, nevertheless, the complete governing equations for a fluid flow were known and available to those people who understood them.

Due to the difficulty of finding solutions to the Navier-Stokes equations, mathematicians and scientists found more simplified means of predicting the motion of fluids. For example, the famous German physicist Hermann von Helmholtz (1821-94) introduced the concept of vorticity and studied the shedding and propagation of vortices in a fluid flow. This rather esoteric-sounding concept provided a new approach for fluid dynamicists to calculate the flow of a fluid. Although not applicable in all cases, this vortex theory expanded the repertoire of fluid dynamicists and later would form the basis for a stunning breakthrough in the calculation of lift.

Frictional force acts between a fluid moving over a solid surface and the surface itself, just as it does when one pushes a book on a table. The surface friction

tends to retard the motion of the fluid; the fluid, in turn, exerts a frictional tug-
ging force on the surface in the direction of the fluid motion. The importance of
this friction force as a contributor to the aerodynamic drag on a body moving
through the air was totally lost on most nineteenth-century investigators. Even
had they realized the significance of friction drag, they had no way of theoretically
predicting the magnitude of this force. An intellectual breakthrough in the eventual understanding of such matters occurred in 1883. Osborne Reynolds, a professor at Owens College (later the University of Manchester) in England, performed some dramatic experiments that identified two types of flow under the
influence of friction: laminar flow, where the streamlines were smooth and regular, and turbulent flow, where the streamlines were tortuous and irregular.
Reynolds studied the conditions under which an initially laminar flow would
become turbulent. Because the frictional force due to a turbulent flow is much
greater than that for a laminar flow, aerodynamicists have a vital interest in the
transition from laminar to turbulent flow. Reynolds established a highly useful
theoretical basis for the analysis of turbulent flow.

The academic community was busy in the nineteenth century developing the
science of fluid dynamics to a rather mature state. These scientists had no interest in helping to transfer this technology to the development of flying machines.
They (and most of the general public) considered those attempting to invent flying machines to be madmen. The great divide that existed between these two com-
munities was not unlike the differences that separated academicians from craftsmen throughout much of the history of science, beginning with the ancient
Greeks. By the middle of the nineteenth century, however, flying machine inventors grew larger in number, and it was time to band together in a more formal
sense. This group of self-educated men, whether they knew it or not, were estab-
lishing a new profession—aeronautical engineering—and they needed some professional identity.

The growing number of powered-flight enthusiasts established technical soci-
eties to formally exchange ideas and publish papers on aeronautics. The first of
these was the Societe Aerostatique et Meteorologique de France, founded in Paris
in 1852. The second, and by far the most important, was the Aeronautical Society
of Great Britain, founded in London in 1866. The first meeting of its council took

place at the residence of the Duke of Argyll on January 12, 1866. The society, in the words of its first honorary secretary, Fred W. Brearey, was "formed to encourage, to observe, to record, and to aid, in proportion as its ability is strengthened by the support of its members." Its establishment created a formal mechanism for technical credibility in aeronautics—proof that the technical aspects of aeronautics were becoming a more accepted field of endeavor. The technical prestige associated with investigations in aeronautics at first came slowly, as is indicated by a statement from the society's fifth annual report in 1870: "Men do not consider the subject of 'aerostation' or 'aviation' to be a real science, but bring forward wild, impracticable, unmechanical, and unmathematical schemes, wasting the time of the Society, and causing us to be looked upon as a laughing stock by an incredulous and skeptical public."[1]

In its first annual report the society published a paper entitled "Aerial Locomotion and the Laws by Which Heavy Bodies Impelled through Air are Sustained." Its author, Francis H. Wenham, delivered the paper on June 27, 1866, making the first statement of an important principle in applied aerodynamics. From extensive studies of birds, he noted "that the swiftest-flying birds possess extremely long and *narrow* wings and the slow, heavy flyers short and wide ones" (emphasis original).[2] He went on to suppose that the wings of flying machines likewise should be long and narrow. Wenham for the first time recognized the advantage of a high-aspect-ratio wing, and he theorized that most of the lift on a wing at moderate angle of attack comes from the front portion of the wing. It followed that the most efficient wing configuration would be that of a number of long, narrow wings arranged one above the next—a multiwing concept. (Horatio Phillips in 1908 employed this design feature—a large number of venetian-blind-shaped wings stacked above each other, in four tandem decks. Phillips was airborne for approximately 500 feet.) Wenham had no experimental or theoretical proof of his statements, and his paper had no mathematics. He had only the birds to go by.

"The steam engine at the present day," declared the Duke of Argyll in the society's third annual report in 1868, "represents the greatest force to be obtained as yet. It is, however, not only a very heavy cumbrous, metallic body in itself, but it requires a large supply of water and fuel to enable it to do its work; these are additional heavy bodies, and the combined weight of the machine altogether precludes the hope that an engine can be so made as to raise or move itself in the air. Still, the absence of the lighter motive power required ought not to stop us from investigating the principle upon which it is to be applied."[3] Here, the duke pointed out the value of engineering research to provide fundamental information on physical principles even though an application of these principles may not yet be at hand.

Total misconceptions sometimes led to correct conclusions. At a meeting of the society in the Crystal Palace, a member offered some illustrations of the rotary motion of a bird's wing. The presenter argued that air never presses on the back of the wing and that the rush of air acts as a sustaining force beneath the curvature of the wing. He concluded, therefore, that the best means of flight is a curved wing. His physical picture of the action of the air on the wing was totally erroneous because the air exerts pressure on the top as well as the bottom of the wing, and the net *imbalance* of this pressure distribution—low pressure on the top and high pressure on the bottom—is the source of lift. Nevertheless, he was correct that an airfoil should be curved. George Cayley had arrived at this conclusion sixty years earlier, based on experimental data.

Another interesting and quite prophetic statement appeared in the annual report in 1868, when another member opined that "a large machine is more likely to succeed than a model." This statement emerged from data showing the "effect produced on one area will not be produced on another" (the first statement of *scale effects* in aerodynamics).[4] An example of scale effects is that friction drag on a small model amounts to a much larger percentage of total drag than on a full-size airplane.

In the same annual report, Wenham and John Stringfellow—in the absence of experience or evidence—stated that the airscrew would prove the "best method of propelling through the air." The writings of the society reflected heated debates over propulsion—whether a propeller or a beating wing would work better. The society took the position that as a means of propulsion "steam was, undoubtedly, the most economical, but in their present state, *gas* would answer better." Here, gas did not mean gasoline but rather an engine where the pistons are driven by some type of hot gas such as might be formed from carbonic acid (as Cayley, too, had thought).

The society summed up its technical frustration in the 1868 report: "With respect to the abstruse question of mechanical flight, it may be stated that we are still ignorant of the rudimentary principles which should form the basis and rules for construction." This is not to say that optimism did not exist. In its report for 1869, the society presented a translation of a French paper that stated: "Science is ripe, industry is ready, everybody is in expectation; the hour of aerial locomotion will soon arrive."[5] Little did they realize that the "arrival" would take another thirty-four years.

During this period, Francis Wenham built the first wind tunnel. In 1870 he designed an experimental device—a long, rectangular, ten-foot duct with an inside cross-section in the shape of a square eighteen inches on each side—for the purpose of measuring lift and drag on small models mounted inside the duct. A steam engine powered a fan that drove air through the duct at about forty miles per hour. Built at Penn's Marine Engineering Works at Greenwich, Wenham's wind tunnel developed an unsteady airstream, making accurate and repeatable measurements virtually impossible. The device had no vanes for guiding the air, making the mean direction of the airstream uncertain. A crude spring balance of complicated design measured lift and drag on a model in the wind tunnel. No pictures or diagrams of Wenham's device exist; there are only written descriptions.

Despite these difficulties, the members of the society welcomed Wenham's results—the first of their kind. The experimental data obtained on flat lifting surfaces showed meaningful lift created at low angle of attack—lift considerably larger than the accompanying drag. In this sense, Wenham confirmed Cayley's earlier whirling arm results, which had retreated into obscurity. Wenham proved that lift-to-drag ratios considerably greater than one can be achieved; this was great and profound news for the aeronautical community. He also found that the center of pressure is near the leading edge, indicating that most of the lift comes from the front part of the wing, thus confirming his earlier supposition in his 1866 paper. These results appeared in the aeronautical society's 1871 report. There is no record of subsequent use of Wenham's wind tunnel to collect additional data.

Wenham continued to be active in the society for another ten years, critiquing work by others, and generally adding a degree of technical competence to the society's meetings. In July 1882, he resigned, the result of a long-standing feud with Fred Brearey. The honorary secretary since 1866, Brearey's lack of technical knowledge and arrogant attitude had been a growing problem. The society collapsed after

Wenham's resignation, publishing its last report in 1893. Four years later Captain B. F. S. Baden-Powell (brother of the founder of the Boy Scouts) rejuvenated the group, and in 1899, Wenham accepted honorary membership and continued to work on various construction techniques for parts of airplanes. When he died in 1908, the society commemorated him as the "father" of aeronautics in his country.

Nearly forty years earlier, in 1872, Horatio F. Phillips, a young man from Streatham, sat in the audience and listened to Francis Wenham give his report on his wind tunnel experiments. Phillips, then twenty-seven, was not impressed. He doubted the quality of the flow in Wenham's wind tunnel and did not like Wenham's total use of *flat* lifting surfaces. Six years later, the Aeronautical Society commissioned two members to expand on Wenham's data, using a large whirling arm with a radius of fourteen feet rather than a wind tunnel. These whirling arm experiments confirmed Wenham's finding that the center of pressure on a flat surface moves closer to the leading edge as the angle of attack is decreased, but they contributed little else.

Again displeased with the quality of the variously obtained experimental data, Phillips in the early 1880s designed and operated an improved wind tunnel. In an effort to avoid the flow imperfections in Wenham's apparatus, Phillips chose a steam injector to suck in air through the entrance of the wind tunnel; he was possibly inspired by railway engines, in which a jet of steam sometimes drove air up the funnel to create a draft for the firebox. Phillips placed the injector in the exact center of the tunnel; upstream of the injector he built a rectangular box with a six-foot length and a square cross-section seventeen inches on each side. Mounted inside the rectangular section was a large block of wood, which reduced the flow area; hence, the region above the block represented a type of "throat region" where the flow velocity was a maximum value—up to sixty feet per second (about forty-one miles per hour). In this region Phillips mounted the aerodynamic test model.

Phillips's wind tunnel results supplied his main contribution to the story of aeronautics. Setting aside Wenham's flat plates and inspired by the shape of birds' wings, he experimented with cambered airfoils with greater curvature over the top than on the bottom. Phillips measured the aerodynamic performance of these airfoils in his wind tunnel and compared it with that of a flat plate also tested in the same tunnel. Although he made minor errors in interpreting his data,[6] the basic finding remains valid: *Cambered airfoils are considerably more efficient lifting shapes*

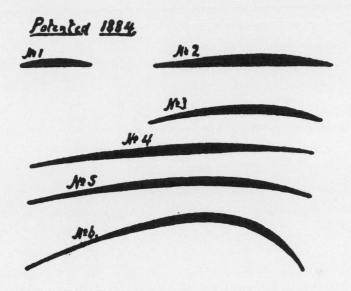

Patent drawing showing Horatio Phillips's cambered (curved) airfoils, 1884. Prior to Phillips's wind tunnel experiments, the wings of flying machines were essentially flat surfaces. Phillips tested a number of cambered airfoils, showing the definite aerodynamic superiority of cambered shapes over the flat plate. Author's collection.

than a flat plate. Phillips's data provided the first *quantitative* proof of this fact, which Cayley had earlier suspected to be true. In 1884, one year *before* he published his results for all to see in the journal *Engineering,* Phillips obtained a patent for the airfoil shapes. A prevailing theory throughout most of the nineteenth century held that the lifting action of an inclined plane moving through the air occurred due to the "impact" of air on the lower surface—a mental picture wrongly reinforced by the Newtonian flow model. Phillips recognized that, when the flow moved over the curved upper surface of the airfoil, the pressure decreased; the lifting action of the airfoil is due to a combination of the lower pressure exerted on the upper surface, and the higher pressure exerted on the lower surface. Although George Cayley had also alluded to this fact, Phillips's widely disseminated results influenced all serious flying machine developers after him.

Phillips later became a developer of flying machines. In 1893 he constructed a large device consisting of fifty wings, each with a span of nineteen feet and a chord (distance between the front and back edges of the wing) of 1.5 inches—creating an ungodly high aspect ratio of 152! (Phillips had clearly adopted Wenham's philosophy of wing design.) With the wings arrayed vertically above a long, cigar-shaped fuselage, the machine resembled a huge venetian blind. Powered by a six horsepower steam engine connected to a single pusher propeller, the unmanned flying machine lifted a total weight of 385 pounds at a speed of forty miles per hour while tethered to a line and constrained to move in a circle of a 628-foot circumference. On the basis of this experiment, Phillips claimed that he definitively proved the viability of cambered airfoils, and he halted testing until the turn of the twentieth century.

TRIAL AND ERROR

Nineteenth-century experiments and information trading—wind tunnels, cambered airfoils, and the proceedings of the Aeronautical Society of Great Britain—counted for technological and intellectual gains. Meanwhile, would-be inventors of the flying machine continued the tradition of trial and error.

Clement Ader built a flying machine that lifted off the ground under its own power. A distinguished French electrical engineer, Ader became interested in powered flight in the midst of a technical career. His approach to the problem of flight did not follow the design characteristics established by Cayley, Stringfellow, and Penaud. Ader did not carry out an extensive series of wind tunnel and whirling arm tests to collect design data. Although familiar with the existing aeronautical literature, he carried out no organized experiments to collect data and patterned his design characteristics after the shape and flight of birds and bats.

Ader built a rather large ornithopter, with a wing span of twenty-six feet and weighing fifty-three pounds in 1872. Though it failed to fly, his obsession with emulating birds only grew stronger. He obtained eagles and large bats from the zoological gardens in Paris and studied their flight in his workshop. He traveled into the interior of Algeria to study large vultures. In 1890 Ader built another large flying machine resembling a bat. Called the *Eole,* this machine had a wingspan of about fifty feet. The *Eole* was a monoplane with a tractor propeller powered by a lightweight yet powerful steam engine that produced twenty horsepower, by far the most

Clement Ader's *Eole.* This machine resembled a large bat, reflecting Ader's feeling that birds provided the best model. A monoplane driven by a powerful, lightweight, twenty-horse-power steam engine turning a four-bladed propeller, the *Eole* took off from the grounds of a chateau at Armainvilliers on October 9, 1890. Piloted by Ader, the machine sustained itself in the air for a distance of 165 feet, becoming the first piloted, powered airplane to take off under its own power. The *Eole,* however, had no flight controls, and Ader had designed the wings to move so as to partially simulate the motion of a bat's wings. The flight of the *Eole* did not satisfy the criteria for a successful, sustained, powered, controlled, piloted heavier-than-air flight. Author's collection.

notable technical contribution Ader's machine made. On October 9, 1890, on the grounds of a chateau at Armainvilliers, Ader piloted the *Eole* as it rolled along the level ground for about ninety feet and then took off under its own power. It sustained itself in the air for another 165 feet before touching down, making it the first piloted heavier-than-air craft to take off under its own power. Yet the *Eole* had no meaningful flight controls. Instead of an elevator Ader tried to reproduce mechanically as many of the motions of a bat's wing as possible (except for propulsion).

Amazingly, Ader obtained funding from the French War Ministry to build a new and bigger airplane. He patterned the new machine, called the *Avion III,* after the *Eole*. It had a wing span of fifty-six feet, weighed about 900 pounds, and was powered by two of his excellent steam engines, each producing twenty horsepower and driving its own tractor propeller. The power loading of Ader's machine was 22.5 pounds per horsepower—plenty of power to do the job. Ader tested the *Avion III* only twice, on October 12 and 14, 1897, on a specially prepared circular track near Versailles. Both tests failed, ending Ader's efforts to fly.

Clement Ader made virtually no contribution to the technology of early flight, nor did he influence anyone to follow in his footsteps. His thinking—so connected with the idea of emulating birds—diverged too far from the mainline development of nineteenth-century aeronautical technology. Nevertheless, even in modern times, he sometimes comes up in French circles as possibly having been the first to successfully fly. By the standard criteria of what constitutes a successful flight, however, Ader does not measure up. (Professor Robin Higham has kindly pointed out to me that Ader did contribute to the aviation world by inspiring the phrase *par avion,* still printed on those little blue labels affixed to airmail letters.)

In the summer of 1894, in England, a huge 8,000-pound flying machine designed and built by Hiram Maxim set a new record—a sustained flight, covering a distance of about 400 feet in the air, at an altitude of two feet! Maxim was another nineteenth-century inventor who dared to work with flying machines. Born in 1840 to a farming family in Sangerville, Maine, he later wrote that he grew up "a poor little bare-headed, bare-footed boy with a pair of drill trousers, frayed out at the bottom, open at the knees, with a patch on the bottom, running wild but very expert on catching fish." After limited schooling, this Tom Sawyer–like figure went to work for a carriage maker, where he acquired mechanical skills, and at the age of twenty he entered his uncle's engineering works at Fitchburg, Massachusetts. Maxim soon began making design changes on the company's gas machines, improving their efficiency. Maxim's star rapidly ascended; by 1873 he was senior partner in the firm of Maxim and Welch, builders of gas and steam engines in New York City, and five years later, as chief engineer, he helped establish the United States Electric Lighting Company. He then obtained the first of his patents dealing with electricity, thus joining those self-educated inventors who were beginning to form the core of the engineering profession.

In August 1881, Maxim departed on the SS *Germanic* to set up an office to represent the interests of the U.S. Electric Lighting Company in London. Within two years, Maxim invented and patented a machine gun design that brought him fame and fortune. In 1884, with the financial backing of "the best men in London" (in Maxim's words), he formed the Maxim Gun Company.

With expense no serious concern, Maxim began to work on a flying machine. Several wealthy gentlemen had asked him about it, and Maxim had taken his usual "can-do" attitude. "The domestic goose is able to fly and why not man be able to do as well as the goose."[7] He estimated that it would take him five years and 100,000 English pounds to accomplish the feat. Maxim rented a large open space—Baldwyns Park at Kent, hired two skilled American mechanics, and built a large hangar.

Prevailing theoretical approaches to flying machine design did not appeal to him, and he certainly did not support the academicians developing the sometimes rather esoteric science of fluid mechanics. Later, in the 1880s, he wrote: "I think we might put down all of their results, add them together, and divide by the number of mathematicians, and then find the average coefficient of error." Maxim called for an approach to manned flight requiring "no more delicate instruments than a carpenter's two foot rule and a grocer's scales."[8]

On the other hand, Maxim did not set out blindly. He began with a series of aerodynamic tests on different airfoil and wing shapes, using a wind tunnel and a whirling arm. Both of these facilities were large; the wind tunnel had a three-foot square test section, and the whirling arm swept out a circle with a sixty-four-foot diameter, with a wire extension at the end of the arm that yielded another circle with a 318-foot diameter. With his whirling arm, using models as large as twenty-five square feet in planform area, Maxim achieved a relative velocity of eighty miles per hour. From his wind tunnel experiments, conducted in a test airstream of forty miles per hour, he concluded that in the angle of attack range of three to seven degrees—the range at which he expected to fly—lift was directly proportional to the angle of attack (not to the square of the sine of the angle of attack as predicted by the Newtonian sine-squared law). He carried out these experiments not for the sake of data collection but "to build a flying machine that would lift itself from the ground." He also noted the variation of aerodynamic force with velocity. "I think it is quite safe to state that the lifting effect of well-made airplanes [i.e., wings], if we do not take into consideration the resistance due to the framework holding them in position increases as the square of their velocity," he wrote, reiterating the velocity-squared law, first confirmed about two and half centuries before. "Double their higher speed and they give four times the lifting effect. The higher the speed, the smaller the angle of the plane, the greater the lifting effect in proportion to the power employed." Maxim realized that, in steady, level flight, lift must always equal weight and that as velocity increases (in order to keep the lift from exceeding the weight) an airplane must fly at a smaller angle of attack.[9]

Maxim conducted propeller tests, experimenting with the influence of blade pitch, shape, and size. He believed a flying machine's propeller, like that of a ship, belonged in the rear. "If the screw is in front the backwash strikes the machine and certainly has a decidedly retarding action." Maxim's logic, based on a seafaring analogy, neglected two considerations. When the propeller pulls from the front (the tractor configuration), its aerodynamic efficiency is enhanced because it faces directly into the undisturbed free stream flow. In contrast, when the propeller pushes from the back (the pusher configuration), the airstream coming into the propeller has been disturbed by flowing over the wings and fuselage, thus compromising the propeller efficiency. Also, with the tractor configuration, the heavy engine rests at the front, moving the airplane's center of gravity forward, improving the stability of the airplane. On the other hand, with a pusher configuration, the flow over the fuselage and wings is smoother compared to the turbulent airflow coming from the wake of a tractor propeller.

Maxim's wind tunnel and whirling arm tests took place during the period 1889–91, after which the design process took over. Maxim envisioned an upper wing with a forty-foot hexagonal section and attached extensions that increased the wing span to 104 feet, resulting in a high-aspect-ratio wing. Originally intending to use a gasoline-powered internal combustion engine (which certainly would have been an advanced feature), Maxim changed to steam power because steam engines were much more mature at that time. In the final design, the boiler weighed a mere 1,000 pounds and fed steam upward to two engines, each driving a large pusher propeller with a diameter of seventeen feet, ten inches. The steam power plant generated 362 horsepower. Maxim's propellers were made of laminated American white pine covered with Irish linen and painted zinc white. The weight of the flying machine was 8,000 pounds, giving it a power loading of twenty-two pounds per horsepower—a very respectable value at that time. For directional control, instead of having a rudder, Maxim planned to use unequal thrust between the two engines. The outer wing panels had dihedral for lateral stability.

On the field at Baldwyns Park, Maxim laid out 1,800 feet of dual railway tracks along which the four-wheel undercarriage would run during takeoff. Maxim aimed to demonstrate that the machine would generate enough lift to leave the ground—and only that. He had no interest in attempting a full-fledged, controlled flight. He attached four extra raised wheels on outriggers that, if the flying machine climbed any higher than two feet, would engage a wooden guard rail directly above them, keeping the machine from climbing any higher.

His test trials with a three-man crew began on July 31, 1894. On the third run on that day, he applied full power and the machine required a rolling distance of only 600 feet before it lifted off the track. The outrigger wheels soon engaged the upper guard rail. At a speed of forty-two miles per hour, after covering a total distance of 1,000 feet along the track, one of the upper guide rails broke, and the machine lifted free for a moment, "giving those on board the sensation of being in a boat." Maxim instantly shut off the steam. In settling back to the ground, the machine suffered some damage, but Maxim, who described himself as a "chronic inventor," had demonstrated that a heavier-than-air flying machine could generate enough lift to leave the ground under its own power. He repaired the machine and, according to reports in the *Times* of London, held a public demonstration in early November. The following July, when he mounted a third spectacle, members of the Aeronautical Society paid him an appreciative visit. Maxim next received notice from his landlords that Baldwyns Park had been sold to the London County Council for a mental home. He also learned that his financial backers had developed cold feet, even though by this point Maxim had spent only 20,000 pounds of the 100,000 he had originally estimated to be the cost. Maxim's machine never flew again.

Maxim's work can best be described as a monumental laboratory experiment, finally proving that a large flying machine could generate enough lift to get off the ground. In *Natural and Artificial Flight,* which Maxim published late in 1908, he described himself as gratified "that all the successful flying machines of today are built on the lines which I had thought out at that time [i.e., 1893-94], and found to be best."[10]

The nineteenth century would-be flying machine inventors that we have met in this chapter—Felix du Temple, Alexander Mozhaiski, Hiram Maxim, and Clement Ader, among others—shared the same approach towards getting into the air. For them, powered flight involved brute force: build an engine powerful enough, attach it to an airframe strong enough, and pull yourself into the air by your bootstraps. They gave little thought to what would happen once you got in the air.

Aerodynamics

Airplanes, like all scientific developments, were no flash of inspiration, but a process of evolution comprising occasional leaps of genius among a sequence of multitudinous little steps.

—Harold Penrose, Aeronautical Historian, 1967

Aerodynamics, propulsion, structures, and flight control, optimized and working together, make a successful airplane. The concept of flight control did not exist in the nineteenth century. Existing civil engineering knowledge influenced structures and the materials used to build them. Propulsion depended on the evolution of steam engines. The flying machine itself, however, drove the study of aerodynamics. The successful design of such machines demands an understanding of aerodynamic lift and drag and a reasonable means to calculate them. By the end of the nineteenth century, through experiments carried out by enthusiasts such as George Cayley, Otto Lilienthal, and Samuel Langley, aerodynamics had evolved into a useful science that provided enough understanding and knowledge to allow the design of a successful airplane.

GEORGE CAYLEY

As we discussed in chapter 2, the story of aerodynamics starts with George Cayley. Because his seminal concept for a flying machine involved a fixed wing that generated lift at a small angle of attack to the airflow, he needed data on the variation of lift with angle of attack. The Newtonian sine-squared law provided the only practical theoretical prediction of lift, and Cayley knew this law to be defective. Frustrated with the lack of data, in 1804 he designed, constructed, and operated the whirling arm device with which he carried out the first meaningful aerodynamic experiments on lifting surfaces. He mounted a flat surface made of paper stretched tightly over a frame at the end of the arm, and tested it at angles of attack from minus three degrees (three degrees *below* the horizontal wind direction) to eighteen degrees. Cayley listed in tabular form his results for aerodynamic force at various angles of attack. A recent study shows that Cayley's measurements were accurate to within 10 percent, remarkable accuracy considering the deficiency of using a whirling arm as a testing device and the lack of sophistication in instrumentation in the early nineteenth century.[1] (He used a simple spring scale to measure aerodynamic force.) Cayley tested a *flat* lifting surface, but he was the first to casually observe that a curved (cambered) airfoil shape produced higher lift than a flat surface at the same angle of attack. Subsequent investigators ignored this important observation until Horatio Phillips rediscovered it by experiment and patented it seventy years later.

Aerodynamic drag also concerned Cayley. He knew that drag needed to be overcome by the thrust of the power plant in order to accelerate the flying machine forward. The importance of low drag was clearly evident. He knew about the value of "streamlining," making an aerodynamic shape long, slender, and gradually tapering off toward the back. He noted: "It has been found by experiment, that the shape of the hinder part of the spindle is of as much importance as that of the front, in diminishing resistance."[2] He carried out measurements of the aerodynamic drag on a flat plate oriented perpendicular to the airstream, which gave a value of 0.0037 for Smeaton's coefficient, in contrast to the value of 0.005 published by Smeaton. Cayley became the first person to recognize the error in Smeaton's value. Leonardo da Vinci had also noted the importance of streamlining, but da Vinci's thinking was lost to future investigators for centuries, and did not play a role in nineteenth century aeronautics. Strangely, Cayley's thoughts about streamlining also had little impact on future investigators. (Streamlining as a means to reduce the aerodynamic drag of airplanes did not come into its own until well into the 1920s.)

Nevertheless, Cayley made momentous contributions to the understanding of aerodynamics and to the design of the flying machine—by comparison all previous investigators pale in Cayley's light. He carried out the first aerodynamic experiments for aeronautical purposes. His work was surpassed only late in the nineteenth century by Otto Lilienthal.

OTTO LILIENTHAL

Otto Lilienthal's neighbors in the fashionable Berlin suburb of Gross-Lichterfield witnessed some unusual goings-on during the year of 1888. In his garden Lilienthal had set up a large whirling arm with a twenty-three-foot diameter, towering fifteen feet above the ground. In the calm, early morning hours, Otto, with the assistance of his younger brother, Gustav, turned on the whirling arm and sent multitudinous model wings with different airfoil shapes swishing through the air. During this year, the Lilienthals carried out thousands of tests in Otto's garden. Their experiments were the culmination of more than twenty-two years of aerodynamic testing in various locations in Germany. This work is one of the most important stories in nineteenth-century aerodynamics.

Starting in 1866, Otto Lilienthal, a German mechanical engineer, carried out a protracted series of aerodynamic measurements of the lift and drag on a variety of different-shaped lifting surfaces. These measurements fell into two categories—those obtained with a whirling arm device and, later, those obtained with fixed models mounted on a post and exposed to the natural wind. The experiments proved beyond the shadow of a doubt the superiority of *cambered airfoils* over flat plates as effective lifting surfaces. Although Horatio Phillips in England tested and patented cambered airfoils, he did not amass a great deal of data to demonstrate their viability. Lilienthal did, and his data made cambered airfoils the norm for successful flying machines after the turn of the century. (Lilienthal did not find out about Phillips's airfoils until he went to file a German patent on cambered airfoils in 1889; he subsequently withdrew his patent application.)

Lilienthal introduced aerodynamic *coefficients* to report the aerodynamic force

Lilienthal's Wing Models

Otto Lilienthal dealt exclusively with thin circular arc airfoil shapes with the
maximum height at the center of the chord (as seen in the illustration repro-
duced from his own drawing.) Camber is a measure of the curvature of the
airfoil. If l denotes the chord length (distance from the front leading edge to
the back trailing edge) and h denotes the maximum height of the airfoil surface above the chord line, then the camber is defined as the ratio of h/l.
Lilienthal experimented with camber values ranging from the slender-arched
value of $1/40$ to the more arched shape of $1/12$. The planform shape of the wing
(shape seen from looking down at the top of the wing) is shown at the bottom
of the illustration. It had curved leading and trailing edges, coming to a point
at each wingtip, with an aspect ratio of 6.48.

Otto Lilienthal, *Birdflight as the Basis of Aviation* (Berlin: R. Gaertners Verlagsbuchhandlung,
1889), p. 65.

data measured on his test models—a seminal contribution. Rather than report-
ing just the raw data—the actual values of the aerodynamic force—Lilienthal
divided his measured forces at various angles of attack by the force measured when
the wing was at ninety degrees angle of attack, that is, when the wing was perpendicular to the flow. Dimensionless values, these ratios are force *coefficients,*
which vary only with the shape of the body and angle of attack; the influence of
Smeaton's coefficient and velocity are simply divided out. Lilienthal published his
results in a book entitled *Der Vogelflug als Grundlage der Fliegekunst* [Birdflight as
the Basis of Aviation] published in 1889. This important book contained by far
the most coherent and useful presentation of aerodynamic data to appear in the
nineteenth century.

Some of Lilienthal's data was independently published and widely disseminated as tables of coefficients. One of them appeared in an article by Octave
Chanute entitled "Sailing Flight" in *The Aeronautical Annual* (1897), published

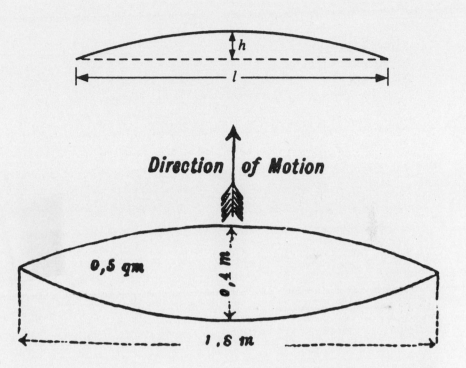

in Boston by James Means; the Wright brothers used the table directly from this source. The data in Lilienthal's table applies only for a circular arc airfoil with camber $^1/_{12}$ and an elliptical wing shape with pointed tips. Unfortunately, this information on the geometry of the airfoil and wing did not appear with the table, leading other investigators (including the Wright brothers) to interpret the data inappropriately and to use it for other geometries. All they knew was that the camber was $^1/_{12}$, as Chanute described when he included the table in his 1897 paper.

Many people assumed that the data in the table came from Lilienthal's whirling arm experiments, but it came from the aerodynamic force measured on stationary airfoils mounted on a stand and placed outdoors in the natural wind. The location was a flat plain, devoid of trees, behind the Charlottenburg Palace in Berlin. The airfoil exposed to the wind experienced lift-and-drag forces measured by vertical and horizontal spring scales. The wind velocity and direction varied—sometimes

Lilienthal's Natural Wind Apparatus

As one of his aerodynamic testing techniques, Otto Lilienthal mounted a wing
model on an outdoor stand and measured the lift and drag as the natural wind
blew over the model. The illustration shows Lilienthal's own drawing of the
natural wind apparatus. At the left is the arrangement for measuring drag: the
wing model *ab* mounted at the top of a long rod experiences drag, *h,* which
causes the rod to pivot in a clockwise direction around the pivot point *m.* The
resulting extension of the spring scale *f* is a measure of the drag force. The drawing on the right shows the arrangement for measuring lift: the rod is mounted
horizontally with the wing model at one end and a counter-weight, *g,* at the
other end. The wing model experiences lift, *v,* which causes the rod to pivot
in a counterclockwise direction around the pivot point, *m.* The resulting extension of the spring scale *f* is a measure of the lift force.

Otto Lilienthal, *Birdflight as the Basis of Aviation* (Berlin: R. Gaertners Verlagsbuchhandlung,
1889), p. 74.

changing many times over the interval of a minute—requiring Lilienthal to simultaneously measure the wind velocity and the aerodynamic force. He accomplished
the wind velocity measurements with an anemometer. With the natural wind device,
Lilienthal repeated the same experiments that he had carried out earlier with a
whirling arm. In his book *Birdflight as the Basis of Aviation* (1889), he plotted graphs
comparing both sets of results. Lilienthal appreciated the uncertainties in whirling
arm data, where the model at the end of the arm constantly moved through the disturbed air from the previous rotation. At the same angles of attack, his natural wind
data gave force values up to 20 percent higher than those obtained with the whirling
arm. Lilienthal properly felt that the natural wind data were more valid, which
explains why they appear in the Lilienthal table. (Recent research shows that the data
in the Lilienthal table is reasonably accurate.[3])

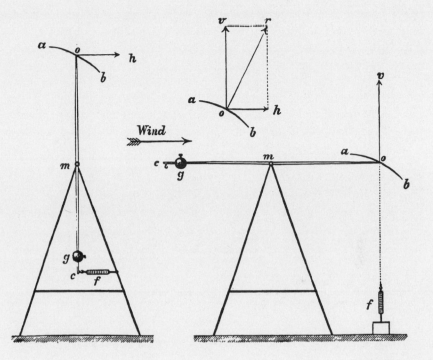

Otto Lilienthal's aerodynamic experiments and data, as published in his 1889 book, represent a significant contribution to the technology of early flight, the most important work since George Cayley's. University educated, with a degree in mechanical engineering, Lilienthal thought differently from the rest of the aeronautical community. He designed and executed his experiments carefully, and he thought in terms of the more technically mature concept of aerodynamic force *coefficients* rather than just the forces themselves. Lilienthal's contributions to aerodynamics were more than enough to ensure his fame and technical reputation. Yet Lilienthal played another exciting and seminal role as the inventor, builder, and flyer of the first practical gliders in history.

As teenagers, Otto and his brother Gustav shared a passion for flight. Starting in 1861, they observed and studied bird flight, constructed wings that they strapped

to their arms, and attempted to fly. After a hiatus caused by their formal education and the Franco-Prussian War, the brothers' aeronautical work began anew on
a more mature level. During an 1873 stay in London, Gustav visited the
Aeronautical Society; both he and Otto became members and quickly learned
about aeronautical progress in Britain. That year, they used their summer vaca-
tion to construct their first whirling arm device; a gymnasium in Berlin provided
the space for their "laboratory." During the same period, they carried out their
first measurements in the natural wind. The Lilienthals did not publish their
results until 1889; Otto later stated the reasons:

> The long delay in publishing our aviation discoveries was nothing more than the nat-
> ural result of the accompanying circumstances. While we were dedicating every hour
> of our free time to the flight question and were already on the trail of the laws which
> would free the solution from its guarded summit, most people in Germany consid-
> ered anyone who would waste his time on such a profitless art to be a fool . . . At that
> time it had just been confirmed once and for all by a particularly learned government-
> appointed commission that man could not fly, which did not particularly lift the spir-
> its of those working on the problem of flight . . . In addition, as young people, com-
> pletely without means, we had to save our money penny by penny, by skipping
> breakfast, in order to be able to carry out our experiments, while at times we were com-
> pletely blocked from our aviation work due to the battle for existence. We were
> absolutely not in any position to produce a good publication of our achievements.

The government-appointed commission, headed by Hermann von Helmholtz, the
most respected scientist in Germany, concluded that human-powered flight was highly
improbable. The wider public, however, interpreted this to include any form of
human flight in a heavier-than-air machine, and it helped to reinforce the already prevailing attitude that would-be inventors of flying machines, greatly misguided, wasted
their time. Aeronautical enthusiasts faced the same deriding public attitude in England
and France. People tried to advance the technology of early flight in spite of the general public, rather than being cheered on in the face of technical adversity. It is no surprise, therefore, that Otto Lilienthal, being a trained mechanical engineer, wanted to
be totally confident of his results before airing them to the skeptical public.

After marrying in 1878, Otto and his wife took up residence in Berlin, where he worked as an engineer and sales representative for the Hoppe machine factory. After being granted a patent for a compact, efficient, low-cost spiral-tube boiler, Otto opened a factory for manufacturing the boiler. (The Lilienthal factory maintained operation until the end of World War I, well after Otto's death.) The factory provided his major source of income for the rest of his life. It grew to employ sixty workers who produced not only boilers but also steam engines, steam heaters, forged pulleys, and chord-sirens for foghorns. Otto's modest income from the factory rose and fell with the economic tides that swept Germany during the latter part of the nineteenth century.

In 1886, Otto and Gustav joined the German Society for Advancement of Airship Travel, which became a public forum for much of their aeronautical work. Founded in 1881 by a few people interested in steerable balloons, the society published the *Journal of Aviation.* By 1889, the society had about a hundred members. Elected in January 1889 to be a member of the technical commission of the society, Otto became secretary in 1891. From 1891, he served on the journal's editorial committee.

Otto and Gustav performed their 1888 aerodynamic experiments to verify their earlier measurements, this time using larger and better instruments. The new results agreed with their earlier whirling arm data from 1873–74. In addition, the Lilienthals repeated their natural wind experiments in an open plain between Teltow, Zehlendorf, and Lichterfelde, again arriving at results that agreed with their earlier findings. Their tests carried on until the brothers finally felt comfortable presenting the results to the public. Otto gave three lectures to the aeronautical society in October 1888 and in February and April 1889. These lectures attracted some attention beyond the membership, including newspaper coverage. Then Otto's book appeared in the fall of 1889. Although 1,000 copies were published by R. Gaertner's company in Berlin, Otto had to pay the printing costs himself. By the time he died seven years later, fewer than 300 copies had been sold. (As late as 1909, copies of the first printing could be found in stores for ten marks. Today, a first edition sells for several thousand marks.)

After publishing his book, Lilienthal carefully redirected his aeronautical activities to the design of flying machines. In the summer of 1889, he built a large wing, 36.3 feet long with a 4.6 foot maximum chord; the planform resembled that of a bird's wing (including pointed wing tips). It had cambered wing sections, reflecting the basic finding of Lilienthal's aerodynamic experiments, and in the middle of the wing was an opening large enough for a human body. Lilienthal never left the ground with this wing; he simply used it to experiment with the strength of the lifting force (which he noted to be considerable) and how the wing might be balanced in the natural wind. A year later, he conducted similar tests with a slightly modified wing. Finally, in a lecture to the society in March 1891, Lilienthal outlined a plan for flying. In his lecture, "On the Theory and Practice of Free Flight," he suggested trying only short, downhill hops. By the late spring, he put his words into practice. Using a glider designed with both a wing and a tail, Lilienthal jumped into the air for short hops in a field in Derwitz, a small farm town beyond Potsdam, accomplishing over a thousand jumps by the end of the summer. In his 1891 annual report to the society, he stated that "in this way I acquired the ability to glide down the gentle slopes of the hill in moderate winds and land at the foot of the hill with no accident of any kind." In this fashion, in the year 1891, Lilienthal became the first person to achieve sustained success with manned glider flights. Later, in 1898, the French aviation pioneer Ferdinand Ferber wrote that "I conceive of the day in 1891 when Lilienthal first sliced fifteen meters through the air as the moment in which humanity learned to fly."[4]

For the years 1892 and 1893, Lilienthal moved his glider experiments to a new practice field on Rauh Hill in Steglitz, only a twenty-minute walk from his house. By this time his gliders had progressed through five design evolutions, each with differences in wingspan, area, and planform, and with some structural changes. Being hang gliders, where the only means of control was by shifting the position of the pilot's body and hence shifting the center of gravity, they could not be made too large or else his body shifting would not be effective. Lilenthal's 1893 glider, with a wingspan of 23.1 feet, a wing area of 152.5 square feet, an aspect ratio of 3.5, and a weight of only forty-four pounds, formed the basis of his first aircraft patent in 1893. The English granted him a patent in 1894 and the Americans in 1895.

In the summer of 1893, Lilienthal also began to fly from some open hills in the Rhinow Mountains, about a hundred kilometers northwest of Berlin. He felt that this terrain was ideally suited for gliding flights and that his machines and flying expertise had matured enough to make it worth the hour-long train ride and sub-

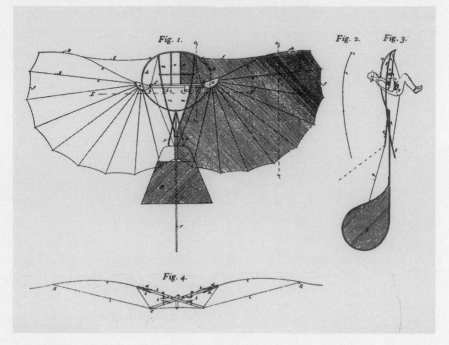

Lilienthal's patent drawing for a glider, 1893. Author's collection.

sequent wagon trip to the flying field. In addition, in 1894 he built a conical-shaped hill in Lichterfelde, a suburb of Berlin; he could fly from the top of this hill irrespective of the wind direction. Built from an existing rubble heap, Lilienthal's Flying Hill reached about fifty feet high. At the apex of the hill he constructed a windowless shed where he stored his gliders. Flying Hill became the main location for his experiments after 1894. In 1932, it became an official monument to Lilienthal. It exists to the present day in the middle of a garden-like park adjacent to a large pond. Seventy-five steps lead to the top of the Flying Hill, providing a panoramic view of the Berlin suburbs. A small pavilion with a ring-shaped roof, under which sits a stone globe on a square basalt base, caps the hill. Stone plaques at the foot of the hill commemorate some of Lilienthal's supporters and helpers. Originally dedicated on August 10, 1932 by the Lord Mayor of Berlin, this memorial stands as a lasting posthumous symbol of Germany's tribute to their first aeronautical pioneer.

At the Flying Hill at Lichterfelde and the chain of hills in the Rhinow Mountains, Lilienthal continued to perfect his flying abilities and his glider designs for the remaining two years before his death. Although he predominantly designed single-wing monoplane gliders, in 1895 he designed and built several biplane gliders, successfully testing them for the first time in August 1895. He continued to fly both types of gliders thereafter. During the course of his experimentation with gliders, from 1891 to 1896, Lilienthal accumulated a great deal of flight time, successfully completing over 2,000 flights using both his monoplane and biplane machines.

Lilienthal moved toward his ultimate objective—the design of an engine-powered, manned flying machine. Strangely enough, he took a technically unsound approach, which stands in stark contrast to his innovative and well-conceived programs of aerodynamic experiments and glider development. Convinced that successful powered flight would emulate bird flight, he focused on ornithopters. Part of his 1893 patent included an ornithopter design with a one-cylinder engine driving the up-and-down wing-beating motion of the outer portion of the wings. A slat-like design, the beating sections of the wings closed during the downstroke (to maximize the "lifting" action) and opened like a fan during the upstroke to minimize resistance. Lilienthal built such a machine in the autumn of 1893 and tested it as a glider at Flying Hill in the spring of 1894. With the engine, the machine weighed over ninety pounds, twice the weight of his unpowered gliders. This machine, with its heavier weight and higher drag, resulted in more steep glide paths and faster landing speeds, making it more difficult for Lilienthal to fly it as a glider. When he attempted to operate the engine, which was powered by compressed carbon dioxide, it froze after a few strokes. These efforts proved totally unsuccessful. In spite of this failure, Lilienthal started to build a second ornithopter during the summer of 1896, using a new engine mounted on a larger airframe with a wing area of more than 218 square feet. His second ornithopter never came to fruition because of Lilienthal's untimely death.

Totally out of character for Lilienthal, a sensible, accomplished, practical mechanical engineer, he pursued blindly the concept of a powered ornithopter. He had knowledge of other work on propeller-driven fixed-wing designs; the German *Journal of Aviation* had many reports about work on such machines. Lilienthal nevertheless preached his advocacy of ornithopters to his colleagues, expressing concern that the

Otto Lilienthal designed and flew the first successful hang gliders. Supporting himself with his arms, his body dangling under the glider, Lilienthal controlled and balanced the machine during flight by swinging his body, hence shifting the center of gravity of the machine plus pilot. Numerous photographs of Lilienthal in flight, such as the one shown here, found wide distribution. Lilienthal became the first human being to be photographed in flight. Between 1891 and 1896, he made over 2,000 successful glider flights. His work contributed greatly to the technology of early flight, and he provided an inspiration to other aviation enthusiasts, including the Wright brothers. Courtesy of the National Air and Space Museum.

slipstream from a propeller would affect the flying qualities of the glider. Since no propeller-driven aircraft had successfully flown by that time, Lilienthal felt vindicated. Propellers, he thought, should only be used as helicopter rotors.

Lilienthal felt very strongly about the social and ethical impact of flying machines. In the summer of 1896 at the Berlin Trade Exposition, for example, he declared: "Cultural progress is to a great degree dependent on whether man will ever succeed in transforming the kingdom of the air into a general, much-used avenue of trade. The boundaries between peoples would then completely lose their significance, because it is impossible to build barriers all the way up into the sky. It is hardly conceivable that customs duties and wars would continue to be possible. The enormous impetus this would give to communication among nations would ultimately blend languages into one world language."[5] (Two world wars in the twentieth century underscore the naiveté of Otto's statements. To his credit, however, the modern world of commercial jumbo-jets has indeed shrunk the world, and the millions of international tourists tend to bring the peoples of different nations closer together.)

Lilienthal also hoped that he could make money in the flying machine business. In a letter to his sister in September 1893, he wrote: "Recently I have pushed the matter of flying more because I have been making such good progress with it. I even think that inventions which have stemmed from it could be used to make money. If the business becomes a hit, we would all be helped by it."[6] Not doing well financially with his boiler factory at the time, he hoped to make flying a sport in which other people would participate, hence providing revenue for himself. He began to receive orders for his gliders, which he priced at 300 marks apiece. By 1895, the price increased to 500 marks for an improved design, namely, his "normal glider," a monoplane design with which he accomplished the majority of his flights. After obtaining his U.S. patent, he attempted to sell it for 5,000 dollars. He enlisted the support of Octave Chanute, who made several inquiries on behalf of Lilienthal. He even advertised in Germany; in the late 1895 edition of Moedebeck's handbook, Lilienthal had a full-page ad in which he offered for sale "Gliding machines for the Practice of Sport Flying." The ad included a photograph of one of his glider flights. All of this came to very little, however. Although he eventually lowered his asking price to 4,000 dollars, he never sold his U.S. patent. Also, he sold only eight copies of his normal glider; the purchasers lived in countries throughout Europe and in the United States. Nikolai Joukowski in Moscow bought one machine; Joukowski later become the most famous Russian aerodynamicist, primarily through his contributions to the circulation theory of lift. In 1896, the newspaper publisher William Randolph Hearst bought the one Lilienthal glider to come to the United States. (After passing through several hands, this glider, after being restored in 1967, now hangs in the Early Flight Gallery at the National Air and Space Museum of the Smithsonian Institution in Washington, D.C.)

Lilienthal's numerous glider flights attracted not only the attention of certain elements of the German public, but along with his book published in 1889 they gained him the curiosity and, later, the respect of the technical aeronautical community throughout the world. The list of technical people who corresponded with Lilienthal, and even those who beat a path to his door, is impressive. Octave Chanute fostered a growing interest in flying machines in the United States. Beginning in the autumn of 1893 and totaling eleven letters over a period of two years, the correspondence between the two aeronautical pioneers discussed technical matters, reports of Lilienthal's glider flights, and the potential U.S. market for selling Lilienthal's patent. Chanute did not read German, and Otto Lilienthal did not read English; Gustav translated Chanute's letters for his brother, and Chanute obtained handwritten translations of Lilienthal's letters. In addition, Lilienthal corresponded with James Means, the Boston editor of *The Aeronautical Annual,* a compilation of papers on various aspects of flight published during 1895–97. This correspondence dealt mainly with several papers Lilienthal published in the *Aeronautical Annual.*

Samuel P. Langley, secretary of the Smithsonian Institution, paid particular attention to Lilienthal. Langley began a series of tests in 1886 and published the results in his book *Experiments in Aerodynamics,* which became available in 1891. Lilienthal owned a copy of this book and knew about Langley's work. Traveling to Europe each summer, Langley delegated one of his assistants to keep him informed about aeronautical progress in Europe. Langley became well aware of Lilienthal's work; he promoted the publishing of Lilienthal's paper "Practical Experiences in Soaring" in the Smithsonian's 1893 annual report. During August 1895, Langley visited Lilienthal in Berlin for two days. Communication was difficult; Langley spoke English and French, but Lilienthal spoke neither language. The first day they met at Lilienthal's factory, and Langley saw the powered ornithopter under construction. On the second day, Langley visited the Flying Hill at Lichterfelde and observed several of Lilienthal's monoplane and biplane glider flights. It was easier to talk on the second day because of the presence of Gustav, who spoke English.

Not overly impressed with what he saw, Langley found the flight demonstrations interesting but felt that not much could be learned from them. In an August 6 letter to his assistant Augustus Herring, he showed more interest in the construction of the Flying Hill than in the glider flights themselves. In this interaction with Lilienthal, Langley showed an attitude that would ultimately contribute to the failure of his own attempts at powered flight. He did not appreciate the value of learning to fly *before* attempting a powered flight; like so many other would-be flying machine inventors, he mainly concentrated on power and lift. In contrast, Lilienthal concentrated on gaining experience flying through the air well before attempting a powered flight, therefore pioneering the philosophy that ultimately led to success.

Later that autumn, Nikolai Joukowski, busy with the development of an aerodynamic laboratory in Moscow, visited Lilienthal. Most impressed, Joukowski observed several of Lilienthal's flights. After returning to Moscow, he gave a lecture before the Society of Friends of the Natural Sciences stating: "The most important invention of recent years in the area of aviation is the flying machine of the German engineer Otto Lilienthal."[7] Clearly, Lilienthal captured the attention of the international aeronautics community, but the German public did not view him as anyone of special importance—just an aviator among many who were attempting what was perceived as impossible. That Lilienthal received very little economic gain from his aeronautical work is another indication of this lack of public stature.

On Sunday morning, August 9, 1896, Lilienthal took an early train from Berlin to Rathenow. There he hired a horse-drawn cab that took him first to Herm's Inn in Stolln, where he picked up his glider (the inn was a convenient storage location for Lilienthal's machines in between weekends in the Rhinow Hills), and then to the foot of Gollenberg Hill. Beautiful weather prevailed—sunny, with a temperature over 68° and a steady breeze from the east at about seven miles per hour. At noon, Lilienthal took off on a long glide from a point high on Gollenberg Hill. Lugging the machine back to the takeoff point, he attempted a second flight. A thermal eddy caught him by surprise, and the glider virtually stopped motionless in the air. Moving his body violently in an effort to control the machine, Lilienthal could not pick up speed. The glider completely stalled and then nosed down, crashing into the ground from a height of fifty feet. Once he was lifted from the glider, it was discovered that Lilienthal had broken his spine.

Conscious while being driven back to Herm's Inn, he called out several times: "I'm still alive, I am Otto Lilienthal from Lichterfelde." A physician from Rhinow attended him at the inn. Much later, Dr. Niendorf described the situation: "I can still see him today, lying on his back with his beautiful, full, blond beard, not remarking about any pain. I basically did not take his injury very seriously, as he could still move both arms well, though he was completely paralyzed from the waist down, a sure sign that his spine must have been broken." Notified by telegram, Gustav rushed to his brother's side in Stolln by early Monday morning. Otto recognized his brother but soon lost consciousness. Immediately transferred to the Bergmann Clinic in Berlin, Otto Lilienthal died at five-thirty that afternoon without regaining consciousness.[8]

A giant in nineteenth-century aeronautics, Otto Lilienthal was second only to George Cayley in his contributions to advancing manned, heavier-than-air flight. Perhaps some of the most reverent words written about Lilienthal came from Wilbur Wright's last article before his untimely death from typhoid fever on May 30, 1912. Writing in the bulletin of the Aero Club of America in September 1912 (published posthumously), Wilbur said:

> Of all the men who attacked the flying problem in the 19th century, Otto Lilienthal was easily the most important. His greatness appeared in every phase of the problem. No one equaled him in power to draw new recruits to the cause; no one equaled him in fullness and dearness of understanding of the principle of flight; no one did so much to convince the world of the advantages of curved wing surfaces; and no one did so much to transfer the problem of human flight to the open air where it belonged. As a scientific investigator none of his contemporaries was his equal.

The last decade of the nineteenth century proved to be a heady and invigorating period of innovation for the technology of early flight. Maxim in England and Ader in France powered themselves off the ground, although in a very halting manner. Lilienthal almost routinely glided smoothly through the skies near Berlin.

Octave Chanute, cautiously optimistic in 1894, suggested that the technology for powered flight was almost at hand: "It will be seen that the mechanical difficulties are very great; but it will be discerned also that none of them can now be said to be insuperable, and that material progress has recently been achieved toward their solution."[9] The technical progress in flying machines enabled Chanute to write and publish a book in 1894 that surveyed and analyzed this progress. The aeronautical community, including the Wright brothers, read this book, aptly titled *Progress in Flying Machines.* Into this vortex of activity stepped Samuel Pierpont Langley, third secretary of the Smithsonian Institution in Washington, D.C.

SAMUEL LANGLEY

Until 1886, the epicenter of applied aerodynamics research lay in western Europe. That year, however, the situation changed; more than 4,000 miles to the west, the seeds for a new center of aerodynamic activity were sown. In August the American Association for the Advancement of Science (AAAS) met in Buffalo, New York. Through the encouragement of Octave Chanute, then a vice president of the AAAS, the subject of aeronautics had a place on the meeting program. In particular, an amateur experimentalist, Israel Lancaster, presented his work on "soaring effigies" of birds—models that he launched in the air. Although not as spectacular as expected, Lancaster's lecture inspired one member of the audience, Samuel Pierpont Langley, then director of the Allegheny Observatory in Pittsburgh. Langley began to think seriously about manned flight.

After his return from Buffalo, Langley obtained the observatory's board of trustees' permission to construct a whirling arm device for aerodynamic experiments. Although the observatory existed for astrophysical observation, and Langley had built his reputation on contributions in astronomy, especially his studies of the sun and sun spots, the board allowed him to construct and operate a major facility for the sole use of obtaining aerodynamic data. Funding for the initial experiments came from a wealthy friend, William Thaw. Completed in September 1887, Langley's whirling arm swept out a circle of sixty-foot diameter, revolving eight feet above the ground, the largest built to date. By comparison, Lilienthal's largest whirling arm had a diameter of twenty-three feet. Both men recognized the importance of having a large diameter so as to minimize centrifugal force effects on the airflow over the lifting surface mounted at the end of the arm. (Centrifugal force acts parallel to the arm and decreases with distance from the center of rotation—the longer the arm, the smaller the centrifugal force.) More importantly, the larger diameter minimized various flow nonuniformities in the wake of the circular motion of the arm. (Hiram Maxim's whirling arm, which came later, was slightly larger—it had a sixty-four-foot diameter. Because Maxim and Langley communicated with each other, it is very likely that Langley influenced Maxim's design.) In 1887, Langley began a series of carefully designed and executed experiments with his whirling arm. These continued for more than four years, resulting in the publication of a book that elevated Langley to world-class status in the contemporary circle of aerodynamic researchers. *Experiments in Aerodynamics,* published in 1891, constituted the first substantive American contribution to the topic. This research and Langley's subsequent work on actual flying machines after he became secretary of the Smithsonian Institution in 1887 broke western Europe's virtual monopoly in aerodynamic experimentation.

Langley left absolutely no doubt about the ultimate goal of his experiments: he intended to explore and uncover the basic physical laws that would scientifically prove the practicability of powered, heavier-than-air flight. Specifically, he wrote in the introduction to *Experiments in Aerodynamics:* "To prevent misapprehension, let me state at the outset that I do not undertake to explain any art of mechanical flight, but to demonstrate experimentally certain propositions in aerodynamics which prove that such flight under proper direction is practicable. This being understood, I may state that these researchers have led to the result that mechanical sustentation of heavier bodies in the air, combined with very great speeds, is not only possible, but within the reach of mechanical means we actually possess."[10]

Before Langley began his whirling arm experiments in 1887, he read the works of Wenham and Phillips, but he did not know about the yet unpublished experiments of Lilienthal. Langley considered himself to be breaking new ground,

essentially filling a technological void. He later wrote of the prevailing situation
and attitudes that existed when he began his experiments:

> The whole subject of mechanical flight was so far from having attracted the general
> attention of physicists or engineers, that it was generally considered to be a field
> fitted rather for the pursuits of the charlatan than for those of the man of science.
> Consequently, he who was bold enough to enter it, found almost none of those
> experimental data which are ready to hand in every recognized and reputable field
> of scientific labor.[11]

Langley published aerodynamic data for flat plates only; his measurements on cambered surfaces can be found in his laboratory records (now in the rare book room of
the National Air and Space Museum). His attention to flat plates is due in part to
his desire to examine the accuracy of the Newtonian sine-squared law, in use since
the eighteenth century to calculate the aerodynamic force on flat plates.

Langley took nothing for granted. He treated the understanding that the aerodynamic force varies as the square of the freestream velocity as a *theoretical* result from
Newtonian theory, to be used only "in the absence of any wholly satisfactory assumption." Langley's doubts are an example of the somewhat dysfunctional status of aerodynamics at that time. A number of previous investigators, including George Cayley,
had discredited the use of the Newtonian sine-squared law for calculating the variation of aerodynamic force with angle of attack, but Langley felt that he had to dip
an oar into the same waters. The variation of aerodynamic force with the *square* of
the velocity had been experimentally proven, beginning two centuries before Langley
with the work of Mariotte and Huygens, and verified by many subsequent investigators, including again George Cayley. Langley accepted this only as a theoretical
result from Newtonian theory—again, taking nothing for granted. Langley's attitude
resulted from his image of aerodynamics at that time as a technical void, plus it was
his nature to be a thorough and exact observational experimentalist.

He showed concern about experimental inaccuracies inherent in the whirling
arm setup. He knew that as the wing model at the end of the arm whirled around
in a circular path, the outer wing tip would see a faster freestream velocity than
the inner tip. (Similarly, if you are standing on the outer rim of a rotating carousel,

you are moving faster than if you are at the inner rim.) This deleterious effect can be minimized by making the model wingspan small and the whirling arm diameter large. Langley addressed another disturbing effect; if the device is housed indoors, the "rotating arm itself sets all the air of the room into slow movement, besides creating eddies which do not promptly dissipate." He felt that "the erection of a large building specifically designed for them (the experiments) was too expensive to be practicable." Therefore, Langley conducted his whirling arm experiments in the open air, and he made every effort to conduct tests only when the outside air was calm. Langley lamented that "these calm days almost never came, and the presence of wind currents continued from the beginning to the end of the experiments, to be a source of delay beyond all anticipation, as well as of frequent failure." In spite of these difficulties and potential sources of error, the nineteenth-century investigators using whirling arms somehow obtained data that was meaningful in its own right in its own time.[12]

Langley made an assumption in the interpretation of his data that had nothing to do with the inadequacies of a whirling arm. He neglected the influence of friction on his aerodynamic force measurements. By the end of the nineteenth century, the calculation of skin friction drag proved unreliable. Even the basic physical mechanism remained a mystery. Constant debate prevailed on whether or not the action of friction between the airflow and the body surface brought the layer of air immediately adjacent to the surface to zero velocity. Langley expressed his (correct) opinion about this in a footnote in *Experiments in Aerodynamics*: "There is now, I believe, substantial agreement in the view that ordinarily there is no slipping of a fluid past the surface of a solid, but that a film of air adheres to the surface, and that the friction experienced is largely the internal friction of the fluid—i.e., the viscosity."[13] Langley made a calculation of skin friction drag using an unreliable friction formula by Clerk Maxwell. He compared the resulting friction drag on a plate at zero angle of attack with the pressure drag on the same plate at ninety degrees angle of attack and concluded the former to be negligible compared to the latter. Comparing friction drag at low angle of attack with pressure drag at high angle of attack is faulty logic, uncharacteristic of Langley. His subsequent intentional neglect of skin friction compromised to some extent his interpretation of his data for plates at small angle of attack.

Langley made measurements of the aerodynamic force on flat plates over a large range of angle of attack, including ninety degrees, that is, with the plate oriented perpendicular to the flow. From these ninety-degree results, he readily calculated values for Smeaton's coefficient—the plural *values* is intentionally used here because the numbers obtained by Langley varied moderately from one test to another. Taking an average value of his measurements, Langley declared that the final value of Smeaton's coefficient was 0.003 when the force is expressed in pounds, the plate area in square feet, and velocity in miles per hour, a far cry from the earlier accepted value of 0.005 obtained from Smeaton's tables.

Langley designed rather sophisticated electromechanical instruments for measuring aerodynamic forces in contrast to the simple weight, pulley, and spring mechanisms that Lilienthal developed for his force measurements. Langley reported his force results in both tabular and graphical form. In the same spirit as Lilienthal, Langley referenced his force measurements to the measured force on the flat plate at ninety-degrees angle of attack; hence, his recorded ratios are simply the aerodynamic force coefficients Lilienthal introduced independently. These coefficients proved to be very useful because an investigator at that time could take the coefficient value for a given angle of attack; multiply it by Smeaton's coefficient (hopefully using the correct value), the area of the plate in question, and the square of the velocity (whatever it may be); and obtain the aerodynamic force on the plate at the given angle of attack and velocity.

Langley carried out a number of experiments measuring the aerodynamic forces in different ways using various electromechanical instruments. A dynamo powered by a small gas engine provided the electricity for his instruments, and the current passed through a system of fixed and brush contacts on the vertical shaft of the whirling arm, then through wires placed along the arm, and out to the instruments at the end. He obtained his first set of measurements with his "result-ant-pressure recorder," directly measuring the magnitude and direction of the resultant aerodynamic force on flat plates at various velocities and angles of attack. Another set of experiments involved the "soaring" of his flat plate models. With the model fixed at a given angle of attack, Langley adjusted the whirling arm velocity to be just right so that the lift on the plate equaled its weight; that is, the plate "soared." At that moment, a specially designed device called the "component-pressure recorder" measured the drag on the plate. Because lift equaled the weight of the plate, Langley obtained lift and drag separately. Langley therefore generated differently and independently two sets of results for the aerodynamic force, one from his resultant-pressure recorder experiments and the other from his soaring experiments using his component-pressure recorder. Both sets of data agreed. This self-consistency between different experimental techniques supported the accuracy of Langley's data although, of course, because he obtained them on a whirling arm they were subject to the uncertainty associated with such a device.

From Langley's point of view, the main value of these measurements was that they proved wrong the Newtonian sine-squared law. At the beginning, Langley explicitly stated that he desired to "investigate the assumption made by Newton that the pressure on the plane varies as the square of the sine of its inclination." Langley obsessed unnecessarily; Cayley and others in Europe had already pointed out that the sine-squared variation did not hold. Nevertheless, Langley added his own data to the existing evidence that the aerodynamic force varied *linearly* with angle of attack at low angles. He commented about the results: "The principle deduction from them is that the sustaining pressure of the air on a 1 foot square, moving at a small angle of inclination to a horizontal path, is many times greater than would result from the formula implicitly given by Newton."[14] He pointed out, for example, that at a five-degree angle of attack, the experimental results gave a resultant force *twenty times* that predicted from the Newtonian sine-squared law. Although not the first to point out such a comparison, Langley considered the results to be especially important because the practicability of sustained, powered flight hinged on them.

Langley's most important aerodynamic contribution came from yet a third series of experiments—he obtained the first *definitive data* showing the aerodynamic superiority of high-aspect-ratio wings. The measurements came from his "plane-dropper apparatus." Mounted vertically at the end of his whirling arm was an iron frame, on which an aluminum falling piece ran up and down on rollers. He attached his flat plate lifting surfaces to this falling piece, with the lifting surface oriented horizontally to the ground. With the lifting surface locked into its highest position, he started the whirling arm, and when the plate came to the desired airspeed, he released it. The plate then proceeded to fall a maximum dis-

Langley's Wing Data

Samuel Langley obtained the first definitive data showing the aerodynamic
advantage of high-aspect-ratio wings. The illustration shows some of his results
for three wings, with wing A having the highest aspect ratio and wing C the
lowest. In Langley's plane-dropper experiments the falling time for a wing
model is a relative measurement of the lift on the model—the higher the lift,
the longer the falling time. In the illustration, the falling time for each wing is
plotted as a function of the airflow velocity. For a given wing, say wing B, as
the airflow velocity increases, the lift increases, and hence the falling time
increases, as shown by the shape of the curve labeled B. More important, however, is the comparison between the curves for the three wings. The curve for
wing A consistently shows a much longer falling time at all velocities. Wing A
has the highest aspect ratio. Clearly, the wing with the higher aspect ratio pro-
duces more lift than the wings with lower aspect ratios.

Samuel Langley, *Experiments in Aerodynamics,* Smithsonian Contributions to Knowledge,
no. 801 (Washington, D.C.: Smithsonian Institution, 1891), p. 32.

tance of four feet (as allowed by the height of the iron frame). The combination
of the horizontal forward motion and the downward falling motion of the plate
through the air resulted in the air velocity being canted slightly upward relative to
the plate, causing the plate to be at an angle of attack to the incoming flow, producing lift. Langley recorded the time it took the plate to fall the four-foot distance. The higher the lift, the longer time it took for the plate to fall the distance
of four feet. A measurement of the required falling time provided an index of the
lifting capacity of the plate. Using this apparatus, Langley tested flat plate wings
of different aspect ratios. His plane-dropper tests clearly showed that the higher
aspect ratio wings took longer to drop than those with lower aspect ratios, proving conclusively that wings with high aspect ratio produce more lift than wings
with low aspect ratio.

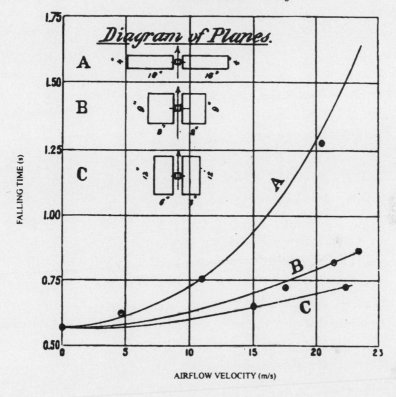

WEIGHT OF MODELS 0 465 kg

FALLING TIME (s)

AIRFLOW VELOCITY (m/s)

The most controversial conclusion Langley made on the basis of his experimental data proved to be the "Langley Power Law," which states that the power required for a vehicle to fly through the air *decreases* as the velocity increases. Langley considered this to be one of his most important contributions. He stated on page one of his *Experiments in Aerodynamics:* "These new experiments (and theory also when reviewed in their light) show that if in such aerial motion, there be given a plane of fixed size and weight, inclined at such an angle, and moved forward at such a speed, that it shall be sustained in horizontal flight, then the more rapid the motion is, the *less* will be the power required to support and advance it. This statement may, I am aware, present an appearance so paradoxical that the reader may ask himself if he has rightly understood it." He repeated this conclusion no less than three other times in his book, twice in italics.[15]

This conclusion flew in the face of intuition, which is why Langley labeled it as paradoxical. At best some contemporaries considered it to be misleading, and Lilienthal and the Wright brothers rejected this conclusion outright. In a meeting of the British Association for the Advancement of Science at Oxford in August 1894, Langley presented a short paper summarizing his work and conclusions. Lord Kelvin and Lord Rayleigh criticized him and took him to task—formidable opposition to say the least. Writers to the present day continue to deride Langley for his power law.

Langley's conclusion, however, was based on his experimental data, and these data *consistently* supported it. A recent study shows Langley's conclusion to be correct for the low air speeds of his experiments.[16] His measurements fell on the back side of the power curve, that regime of flight at very low speeds where the power required for steady, level flight decreases as the airplane's velocity increases. At the higher velocities associated with normal flight, this trend reverses, and more power is required to fly faster. If Langley had been able to conduct his experiments at higher air speeds, he would have observed this reversal, and the Langley Power Law would have been different. (See Appendix A for a more detailed explanation.)

Like Lilienthal, Langley designed and flew flying machines although he worked first and foremost as a scientist. He held no formal college degree, but in the earlier years of his life he became interested in observational astronomy. In 1866 he became professor of physics and director of the Allegheny Observatory at the Western University of Pennsylvania (now the University of Pittsburgh). Over the next twenty years, Langley carved out a distinguished reputation as director of the observatory and as an observational astronomer. His lack of mathematical background pointed him in the direction of experimental rather than theoretical astronomy, and this same experimental emphasis later dominated his aeronautical work. On the practical side, he set up a source of income for the observatory by providing the exact time of day to the railroads; he calculated the time from star observations, and then telegraphed the results to his customers twice a day. On the scientific side, he specialized in studying the sun, especially sun spots and the energy produced by the sun. In the late 1870s, he developed the bolometer, an instrument to measure the spectral variation of the sun's energy incident on the earth; he widely published results obtained with this instrument. In 1884 Langley organized an expedition to Mount Whitney in the Sierra Nevada of eastern California to measure the heat-absorbing characteristics of the atmosphere; two years later he published a widely used 239-page report on the results. From his data he determined a value for the solar constant (a measure of the amount of energy reaching the earth's atmosphere). This work brought him lavish praise from his scientific colleagues and solidified his international reputation. He reached the pinnacle of his scientific reputation in 1886 when both the Royal Society in London and the American Academy of Arts and Sciences awarded him the Rumford medals, and the National Academy of Sciences honored him with the Henry Draper medal.

At this time, Langley accepted the job of secretary of the Smithsonian Institution. He also began a totally new scientific career—the study of mechanical flight. In 1887 he moved permanently to Washington, but under his direction the aerodynamic work using the whirling arm continued at the Allegheny Observatory. He maintained close supervision by mail and telegraph. By the time he completed the whirling arm flat plate tests in 1890, he had amassed a bulk of aerodynamic data that was unparalleled in the United States and that, in sheer bulk and variety, exceeded any previously obtained in Europe. This data convinced Langley of the practicality of mechanical flight.

Langley originally wanted to demonstrate the physical laws that would prove the practicality of mechanical flight. At the end of *Experiments in Aerodynamics,* he summarized the impact of his data: "The most important general influence from these experiments, as a whole, is that, so far as the mere power to sustain heavy bodies in the air by mechanical flight goes, *such mechanical flight is possible with engines we now possess.*"[17] He began to realize, however, that to convince the rest of the world that his conclusion was correct, he had to do more than conduct laboratory experiments. Indeed, during Langley's participation in the August 1894 meeting of the British Association for the Advancement of Science at Oxford, Lord Rayleigh commented that the ultimate proof of the validity of Langley's aerodynamic data would be "if he . . . succeeded in doing it [flying] he would be [proven] right."[18] Rayleigh's comment supported activities that Langley already had undertaken—activities aimed at the design, construction, and flying of a heavier-than-air machine. Langley's goal was clear—he had to build a successful flying machine.

Langley drew on experience obtained with small, rubber-powered aircraft models. This work had commenced in April 1887 at the Allegheny Observatory and continued for four years, mostly in Washington. He made the earlier models from pine; later, he replaced those wooden frames with light metal tubes, which proved too heavy, and still later with shellacked paper tubes, which proved to have the best strength-to-weight ratio. He made the wings by stretching paper over a supporting frame. Over the course of these flying model tests, Langley experimented with nearly a hundred different configurations. He described these configurations as "some with two propellers, some with one, some with one propeller in front and one behind; some with plane, some with curved wings; some with single, some with superimposed wings (biplanes); some with two pairs of wings, one preceding and one following (tandem wings); some with the Penaud tail; and some with other forms."[19] Observers witnessed quite a sight seeing these rubber-powered model airplanes being tossed out the north window of the dome of the Allegheny Observatory, and later flying inside the upper hall of the original Smithsonian building.

"The objects of these experiments," he wrote, "were to find the practical conditions of equilibrium and of horizontal flight." Langley's efforts with these model aircraft, however, proved unproductive, and the mixed results were not definitive. He later wrote:

The difficulties of these long-contained early experiments were enhanced by the ever-present difficulty which continued through later ones, that it was almost impossible to build the model light enough to enable it to fly, and at the same time strong enough to withstand the strains which flight imposed upon it. The models were broke up by their falls after a few flights, and had to be continuously renewed, while owing to the slightness of their construction, the conditions of observation could not be exactly repeated; and these flights themselves, as has already been stated, were so brief in time (usually less than six seconds), so limited in extent (usually less than twenty metres), and so wholly capricious and erratic, owing to the nature of the rubbermotor and other causes, that very many experiments were insufficient to eliminate these causes of mal-observation.

Finally, Langley gave up with the rubber-powered models, stating that: "The final results . . . were not such as to give information proportioned to their trouble and cost, and it was decided to commence experiments with a steam-driven aerodrome on a large scale."[20]

Langley embarked on the next stage of his aeronautical work—the development of steam-powered flying machines. Wishing to give his machines a name derived from the classical languages, Langley in December 1890 consulted a scholar who suggested the Greek word *aerodromoi*, meaning "air runner." Henceforth, he called all his flying machines, including his rubber-powered models, *aerodromes*. Strictly translated, the word *aerodrome* means a *place* from which a machine would fly, rather than the machine itself. The term did not stick, being rarely used by anyone after Langley.

At the time, Langley felt that steam was the power mode of choice. "In November 1891," he wrote, "I commenced the construction of the engines and the design of the hull of a steam-driven aerodrome, which was intended to supplement the experiments given in *Aerodynamics* by others made under the construction of actual flight."[21] Over the next four years, Langley built seven such aerodromes, numbering them consecutively from 0 to 6. He quickly abandoned numbers 0 to 3 because they proved to be too heavy and underpowered. The lessons learned with numbers 0 to 3, however, led to more successful designs for numbers 4 to 6. Of these, aerodrome number 5 achieved the highest degree of flying success. Representative of all of Langley's aerodromes, number 5 had a tandem wing design (one wing behind the other)—a feature that stemmed from the rubber-powered model tests. Both wings had the same shape and size. The planform was rectangular with a relatively high aspect ratio of five; here Langley used to advantage one of the important conclusions from his whirling arm experiments. Both wings had a wing span of 13.1 feet, and the total sustaining wing area, including both wings, was 68.9 square feet. The aerodrome weighed 26 pounds, yielding a wing loading (weight divided by wing area) of 0.38 pounds per square foot. The airfoil shape had a high camber, in the ratio of one to twelve (the same as the airfoil in the Lilienthal tables), with the maximum height of the camber located 23.8 percent of the chord length from the front edge of the wing (for Lilenthal's airfoil, the maximum height was at 50 percent of the chord). Although all the aerodynamic data that Langley published in 1891 pertained to flat plates, a few years later his assistants experimented with cambered airfoils. Never publishing the data, Langley nevertheless chose to use a cambered airfoil on his aerodromes, following in the footsteps of Otto Lilienthal.

Langley enjoyed the best intellectual and scientific circles in Washington. Alexander Graham Bell, one of his closest friends and supporters, would later contribute to aeronautics by forming the effective Aerial Experiment Association in 1907. Another close friend, Albert Zahm, directed the Department of Physics and Mechanics at Catholic University, located just three and a half miles north of the Smithsonian. Zahm built the first aerodynamic laboratory in an American university. Langley lived at the Cosmos Club, still today one of the most prestigious addresses in Washington, and he remained a bachelor all his life.

Langley achieved glorious success in 1896. To launch the aerodromes, he chose the unique arrangement of a catapult mounted on top of a houseboat in the middle of the Potomac river. He later wrote: "As the end of the year 1892 approached and with it the completion of an aerodrome of large size which had to be started upon its flight in some way, the method and place of launching it pressed for decision. One thing at least seemed clear. In the present stage of experiment, it was desirable that the aerodrome should—if it must fall—fall into water where it would suffer little injury and be readily recovered, rather than anywhere on land, where it would almost certainly be badly damaged." In regard to the aerodrome itself, Langley felt challenged by four big problems—power, weight, structural strength, and vehicle stability. To address the power problem, Langley built a small steam engine producing a maximum of one horsepower. (He calculated the power required to attain the theoretical soaring velocity of twenty-four feet per second to be 0.35 horsepower.) The strength-to-weight ratio is also an important consideration in aircraft structures. Langley ultimately solved this problem for the steam-powered aerodromes, but not without effort and frustration. He commented on some abortive tests of aerodromes number 4 and 5 in 1894: "Observations of the movement of the two aerodromes through the air, as seen by the writer from the shore, seemed to show, however, that the wings did not remain in their original form, but that at the moment of launching there was a sudden flexure and distortion due to the upward pressure of the air." That is, the wing bent upward along the span and twisted so that airfoil sections at different locations inclined at the wrong angle of attack. Langley successfully fixed this problem for the small aerodromes, but his observation anticipated the disasters to come with his full scale aerodrome in 1903.[22]

Most of all, Langley obsessed over the need for inherent stability of the aerodrome in flight. He understood the basic principles of static stability. To obtain lateral stability, he designed his wings with a substantial dihedral angle of fifteen degrees. Langley had difficulty obtaining longitudinal stability because he could not predict with any certainty the location of the center of pressure for the wings, let alone that for the complete aerodrome. His flat plate data obtained with the whirling arm did not help—the center-of-pressure variation for a cambered airfoil differs from that of a flat plate. Langley also worried about the mutual interaction of the tandem-wing arrangement and the effect of the propeller slipstream on center-of-pressure location. He faced the engineering problem of having to design and build a vehicle without complete information. When Langley finally prepared to fly his aerodromes, he adjusted the location of the center of gravity by moving various components (the tail, the hull, etc.), and he adjusted the center-of-pressure location by changing the tail inclination angle as well as the angle made by the wing to the fuselage. Through his system of structural guy wires, he set the inclination angle of the wing tip to a different inclination angle than that at the wing root, typically eight degrees at the root and twenty degrees at the tip. After trial and error, Langley found the proper arrangement that provided longitudinal stability and balance for horizontal flight.

On May 6, 1896, after three years of frustrating failures, Langley finally met with success. He moored the houseboat with the aerodrome and catapult on the west side of Chopawamsic Island, a small dollop of land off the western bank of the Potomac River, near Quantico, Virginia. At this location, about fifty miles below Washington, the river afforded a wide expanse of water for the aerodrome to fly over. Delayed by high winds in the morning, the launch crew, with Langley and Alexander Graham Bell in attendance, attempted to fly aerodrome number 6 in the early afternoon, but the catapult mechanism fouled the aerodrome, breaking the left wing and plunging it into the water. After recovering that aerodrome, the crew mounted Aerodrome Number 5 on the catapult. At 3:05 P.M., from a height of twenty feet above the water, they launched number 5 into a gentle breeze. After slowly descending three or four feet, the aerodrome began to rise and to circle to the right, executing a spiral path at a height of about seventy to a hundred feet above the river. After about a minute and twenty seconds, the engine began to run out of steam, the propeller slowed down, and the aerodrome began to descend. By the time the machine settled gently in the water, it had been airborne for a minute and a half, and had covered a distance of 3,300 feet. Immediately recovering the aerodrome from the water, they launched it for a second time at 5:10 P.M. The ensuing flight performed very much like the first; aerodrome number 5 remained aloft for one and a half minutes, this time covering a distance of 2,300 feet, the difference being due to a different wind pattern than earlier in the afternoon. Langley and Bell felt justifiably elated.

That afternoon Langley achieved the most important advance and supplied the most dramatic event in powered flight to that time. Langley's assistant, Charles Manly later wrote: "Just what these flights meant to Mr. Langley can be readily understood. They meant success! For the first time in the history of the world a device produced by man had actually flown through the air, and had preserved its equilibrium without the aid of a guiding human intelligence. Not only had this device flown, but it had been given a second trial and had again flown and had demonstrated that the result obtained in the first test was no mere accident."[23] Samuel Langley had achieved the first successful flight of an engine-powered, heavier-than-air flying machine in history.

Langley did not lose any time in spreading the news. On May 26, he gave a short paper on his success to the French Academy of Science in Paris, emphasizing that his aerodrome flights proved the technical feasibility of heavier-than-air powered flight and that such flight could be accomplished with existing technology. For reinforcement, Langley also attached to his report a letter from Alexander Graham Bell describing in detail what he had observed. Bell ended his letter: "It seemed to me that no one could have witnessed these experiments without being convinced that the possibility of mechanical flight had been demonstrated."[24] This had been Langley's goal all along, and who more prestigious than Bell himself could have confirmed the realization of this goal.

Considerable progress in the technology of flight occurred during the nineteenth century, beginning with the seminal work of Cayley and culminating in the experiments and flying machines of Lilienthal and Langley. This progress set the stage for the invention of the successful piloted airplane.

Technology

Collecting Data

I am an enthusiast, but not a crank in the sense that I have some pet theories as to the construction of a flying machine. I wish to avail myself of all that is already known and then if possible add my mite to help on the future worker who will attain final success.

—Wilbur Wright, 1899, in a letter to the Smithsonian Institution

Wilbur and Orville Wright showed interest in mechanical flight as early as childhood when their father, Milton Wright, brought home in 1878 a small toy helicopter designed by the Frenchman Alphonse Penaud; Wilbur was eleven and Orville, seven. Alexander Graham Bell had filed for a patent for the telephone in 1876, and Thomas Edison had invented the phonograph in 1877. The Wrights' interest in flight was renewed when they read about Otto Lilienthal's successful gliding experiences in 1896. Orville recollected much later: "His death (Lilienthal's) a few months later while making a glide off the hill increased our interest in the subject, and we began looking for books pertaining to flight." In 1899, Wilbur wrote a letter to the Smithsonian requesting information about publications on flight—the first official record of the Wright brothers' serious intent to take up the challenge of mechanical flight. Little did Wilbur know that they were the ones destined to "attain final success."[1]

Why did the Wrights achieve success in designing and flying the world's first practical airplane whereas so many others before them, some with impressive technical qualifications, failed? What aspects of the Wrights' technology led to success? Did they make mistakes along the way? How much of the existing state of the art in aeronautics did the Wrights utilize; indeed, how much did they even know? The period of development of the early technology of flight ends with the advances made by Orville and Wilbur Wright. From their work blossomed the modern twentieth-century profession of aeronautical engineering.

WHAT THE WRIGHTS KNEW

The Wrights, of course, inherited a substantial body of flight technology, from it launching their own research and development. On one hand, much of the existing state of the art, incomplete to say the least, led to confusing, contradictory, and misleading ideas. On the other hand, some fundamental principles and useful design features did emerge prior to the Wrights' work, particularly during the nineteenth century.

Aeronautical technology in 1899 comprised four general categories: theory, experimental data, design considerations, and philosophy. Theory, the least understood and most conflicting aspect of the state of the art, proved to be of no real use to the Wrights or anybody else at that time. Theory and experiment alike showed that the aerodynamic force acting on a flight vehicle depended on the air density multiplied by the size (reference area) of the machine and, most importantly, multiplied by the square of the flight velocity. Understanding performance depended on this knowledge, which the inventors had well in hand by 1899. Academicians had derived the fundamental equations of motion of fluid dynamics—the Euler equations for an inviscid flow (flow neglecting the influence of friction) and the Navier-Stokes equations for a viscous flow (flow with friction). A powerful intellectual accomplishment, these equations contained and described the essential physics of most fluid flows, including the aerodynamic flow over flying machines. Highly nonlinear partial differential equations that could not be solved, these equations proved worthless to the Wright brothers and every other would-be flying machine inventor. Neither mathematicians nor academicians, the

Wrights, with their high-school educations, fitted squarely in the category of craftsmen who were struggling to design things and make them work. No proof exists that the Wrights even saw the Euler and Navier-Stokes equations during their inventive period, let alone understood them. Newtonian theory for the aerodynamic force on an inclined surface predicted that the force varied as the sine-squared of the angle of attack—wrong and totally misleading. Fortunately, by 1899 the application of Newtonian theory for this purpose had been debunked by such respected men as George Cayley, Lilienthal, and Samuel Langley, who obtained experimental data showing that the aerodynamic force varied *linearly* with angle of attack, at least for the small to moderate angles of attack associated with airplane flight. The Wrights knew enough to never touch Newtonian theory.

In contrast to the paucity of useful theoretical results in 1899, there was a substantial and growing bank of experimental data. Smeaton's coefficient had been used since the end of the eighteenth century to predict the aerodynamic force on a surface mounted perpendicular to a flow. Smeaton quoted a value of 0.005 for this coefficient when the force is given in pounds, the area in square feet, and the airspeed in miles per hour. As early as 1809, Cayley published data indicating the value of Smeaton's coefficient to be closer to 0.0037, and near the end of the nineteenth century Langley obtained several careful measurements proving the value to be close to 0.003. Uncertainty about Smeaton's coefficient nevertheless still reigned in 1899, particularly in the minds of the Wright brothers. Lilienthal provided by far the best data on the lift-and-drag coefficients on cambered airfoils as a function of angle of attack, and the Wrights seized upon this data. Langley also obtained an extensive set of data for the variation of aerodynamic force coefficients as a function of angle of attack, but it applied exclusively to flat plates, not cambered airfoils, making the data less valuable to would-be designers of flying machines. The data convinced Langley that it took *less* power to fly faster. The Langley Power Law seemed counterintuitive to almost everybody, including Langley himself. Langley's conclusion applied correctly to the low range of flight velocities at which he conducted his whirling arm experiments—the data were all on the back side of the power curve. At the time, however, nobody understood this phenomenon. As a result of these deficiencies, real and perceived, most flying machine enthusiasts looked elsewhere for data. This proved unfortunate because

buried in Langley's data could be found the first definitive measurements showing the benefits of high-aspect-ratio wings—a very important design feature for
airplanes that went virtually unnoticed in 1899.

From the trial and error of would-be inventors, certain design features of airplanes became accepted as the norm, although no powered heavier-than-air,
piloted machine had successfully flown by 1899. These design features evolved with
experience rather than from the application of some well-known understanding
of the physics of flight; such understanding simply did not yet exist. Most serious nineteenth-century flying machine enthusiasts accepted the concept of the
airplane as first set forth by Cayley in 1799—a machine with fixed wings, fuselage,
and tail, and a propulsive mechanism (separate from the lift-generating mechanism) with enough power to propel the machine forward at a high enough velocity that the wings, inclined at a small angle to the flow, could generate enough lift
to sustain the machine in the air. By 1899, enthusiasts took this design approach
to mechanical flight for granted, and of course the Wright brothers completely
bought into it. Based on Horatio Phillips's work as well as Lilienthal's definitive
aerodynamic experiments, the cambered airfoil became the accepted general shape
for a wing section. A flat surface—a flat plate airfoil—would work, but performed
less efficiently (higher drag for a given lift). Even Langley, who carried out definitive experiments on flat plates, used cambered airfoils on all his aerodromes. By
1899, flying machine inventors, including the Wrights, held a clear preference for
cambered airfoils.

Penaud in the mid-nineteenth century first fully understood the nature of
inherent stability, initially discussed and employed by Cayley. Penaud employed
a negative tail-setting angle to obtain proper longitudinal balance and stability.
Lilienthal for his gliders and Langley for his aerodromes made good use of the
"Penaud tail." The need for inherent stability in a flying machine became an obsession with Langley and with Octave Chanute, who designed some relatively successful hang gliders from 1896 to 1904. By 1899, inventors considered the design
of a flying machine without inherent stability unthinkable. Various inventors
explored the use of monoplanes (single wing), biplanes (two wings, one set above
the other), triplanes (three wings set above each other), and multiplanes. Langley
used two wings, but he put them in tandem. The argument against monoplanes

and for biplanes hinged on structural concerns. One wing made large enough to generate the requisite amount of lift might be structurally unsound and would break under the massive air loads. By sharing the requisite wing area over two or more wings, they could be made smaller and stronger. Late in the nineteenth century Chanute became a strong proponent of the biplane configuration in combination with the Pratt truss concept of using diagonal braces as load-carrying members between the two wings; Chanute used two crossed steel wires between the wings as diagonal bracing members. Such a trussed biplane configuration had amazing structural strength and rigidity. By 1899, however, people had no clear preference for one or the other wing configuration. Due to his status as a highly respected civil engineer, Chanute's choice of the trussed biplane configuration carried weight with some of the fledgling aeronautical community. The Wrights chose the design for all their flying machines.

Inventors had no appreciation of and no means for lateral control, that is, control in regard to rolling motion, a fundamental concept completely missing from state-of-the-art aeronautical technology in 1899. In contrast, they did appreciate lateral stability, and such inventors as Penaud, Lilienthal, and Langley insured it by the use of wing dihedral. Early flying machine design also missed the basic concept that a flying machine is a system and that all components of the system must work together before meaningful success can be achieved. A good airplane is a successful integration of aerodynamics, propulsion, structures, and flight dynamics. In 1899, flying machine design treated these elements somewhat independently; inventors showed little consciousness that a given flying machine is only as strong as its *weakest* component.

Some aspects of a philosophical nature ran through the early technology of flight as it existed in 1899, and these were not lost on the Wright brothers. The "airman's" philosophy, first established by Lilienthal, propounded the need to first get into the air and learn to fly a machine before putting an engine on it. The opposite "chauffeur's" philosophy still lived, however; people tried to claw their way into the air by brute force, with little attention given to how they would actually fly the machine once they got off the ground. Langley took this approach, as did Hiram Maxim, perhaps the most blatant chauffeur of them all. From the beginning, the Wright brothers followed the airman's philosophy. In addition, proponents of mechanical flight clung to the faith that it would someday become a reality—a faith based on a common philosophy of man's eventual harnessing of the forces of nature. Cayley, Maxim, Lilienthal, Chanute, and Langley all shared this belief. Langley, as the secretary of the Smithsonian, was the most prestigious scientist in the United States. If he believed in mechanical flight and actively worked to achieve it, then surely there was hope for others, such as the Wright brothers, on the same quest.

Before Wilbur drafted his letter to the Smithsonian, the Wrights had little familiarity with the existing state of the art. They had read some of the popular literature, such as articles about Lilienthal's glider flights in the widely circulated McClure's Magazine. The Wrights actually knew of Lilienthal as early as 1890, when they published two short articles about him in a newspaper they printed and sold for a brief time. In general, however, the Wrights were in the dark about the *technology* of manned flight. Curiously, Wilbur began his letter, "I have been interested in the problem of mechanical and human flight ever since as a boy I constructed a number of bats of various sizes after the style of Cayley's and Penaud's machines." It is not clear where he obtained information about these inventors' basic configurations. The Wrights had not read Octave Chanute's *Progress in Flying Machines,* published in 1894. A well-known civil engineer, Chanute lived in Chicago, less than 250 miles from the Wrights' home in Dayton, Ohio. His book provided a thorough technical review of all flying machines developed to that time. By no means a best seller, the book did not enter the sphere of the Wrights' everyday activities.

The Smithsonian replied quickly to Wilbur's request for "such papers as the Smithsonian Institution has published on this subject (mechanical flight), and if possible a list of other works in print in the English language." Assistant secretary Richard Rathburn replied on June 2, 1899, enclosing a list of publications. He included several of the most important contemporary sources: Chanute's *Progress in Flying Machines,* Langley's *Experiments in Aerodynamics,* and Means's *The Aeronautical Annual* for 1895, 1896, and 1897. Rathburn also sent copies of four Smithsonian pamphlets, including one written by Langley and another by Lilienthal. Wilbur wrote back on June 14, thanking the Smithsonian and enclosing a dollar for the purchase of *Experiments in Aerodynamics.* The brothers quickly immersed themselves in the existing literature. They obtained a copy of *Progress in Flying Machines* in 1899 and studied it carefully. We have no precise record of the full extent of what they read and when, nor of their detailed reactions at that time. However, in 1920, when reflecting on this period, Orville stated: "On reading the different works on the subject we were much impressed with the great number of people who had given thought to it (mechanical flight), among some of the greatest minds the world has produced . . . After reading the pamphlets sent to us by the Smithsonian we became highly enthusiastic with the idea of gliding as a sport." During the summer of 1899, the Wrights made a quantum leap in their familiarity with the existing aeronautical literature, and they became anxious to apply their newfound knowledge.

THINKING ABOUT IMPROVEMENTS

Products of the post–Civil War age, Wilbur was born near Millville, Indiana, on April 16, 1867, and Orville in Dayton, Ohio, on August 19, 1871. They had five siblings—two older brothers, Reuchlin and Lorin; twins who died in infancy; and a younger sister, Katharine. Their father Milton became an active minister, administrator, and (later) bishop in the United Brethren church; he frequently traveled on business. Much of the staunchly ethical and unbending nature of Wilbur's and Orville's personalities can be attributed to that of their father. Shy, quiet, and scholarly, their mother, Susan Koerner Wright, attended Hartsville College, but left three months before graduation. She excelled at school and was gifted with considerable mechanical ability. After her marriage, she made simple household appliances for herself and toys for the boys, who came to her for advice on mechanical matters. Susan died of tuberculosis in 1889; it was a tragic loss for the Wrights.

Katharine, who had looked after the family during her mother's illness, continued to do so after Susan's death. At that time, four people lived in the Wright's home at 7 Hawthorn Street in Dayton—Milton, Wilbur, Orville, and Katharine. They fused into an extremely tight group, sharing all problems and collectively working out solutions. In spite of her family responsibilities, Katharine graduated from Oberlin College, near Cleveland, in 1898, and took a job teaching classics at Steele High School in Dayton. (The closeness of the Wright family was

destroyed in 1926 when Katharine married Henry Haskell, editor and part owner
of the Kansas City *Star.* Orville considered this a rejection—Wilbur had been dead
for fourteen years—and he completely cut off all contact with Katharine. In his
mind, Katharine had broken the sacred pact forged after their mother's death.
Even when Katharine lay dying of pneumonia in 1929, Orville steadfastly refused
to visit her until the very end when he traveled to her bedside at the insistence of
his older brother Lorin.)

Wilbur attended high school in Richmond, Indiana, where the family then
lived. An accomplished scholar, he took courses in Greek, Latin, geometry, natural philosophy, geology, and composition, earning grades well above 90 percent.
In addition, he excelled as an athlete, especially in gymnastics. Because of his
scholastic achievement, Milton and Susan considered sending him to Yale. Just
before Wilbur was to graduate in June 1884, compelling political reasons dealing
with church business caused Milton to abruptly move the family to Dayton, pre-
venting Wilbur from completing the courses required for graduation. He never
received a high school diploma, and he never attended Yale. He continued to be
a scholar, however, ultimately becoming better read than most college graduates.
Orville attended Central High School in Dayton. Unlike Wilbur, Orville did not
stand out as a student, but he managed. His grades fell in the 70 to 90 percent
range. Just before he was to return for his senior year, his mother died. He had
already decided not to return, mainly because he had taken several special non-
credit advanced courses in his junior year, falling several credits shy of the number needed to graduate with his classmates. Instead, he pursued work as a printer,
including building his own printing press, but his venture proved unsuccessful.

Wilbur and Orville became bicycle enthusiasts in 1892, competing in local
races. That year they opened a bicycle shop on West Third Street in Dayton. First,
they only sold and repaired bicycles. By 1895, however, they manufactured and
sold their own line of bicycles—a profitable business. A major mode of transportation, the bicycle reached its height of popularity in the 1890s; more than three
hundred companies in the United States alone manufactured over a million bicycles a year. Stimulating work using structural design and lightweight but strong
materials, bicycles were a high-tech business at the end of the nineteenth century. The Wright brothers did not, however, adopt the mass production techniques

Origin of the Bicycle

A mid-nineteenth-century invention, the origin of the bicycle traces back to a Scottish blacksmith, Kirkpatrick Macmillan, who constructed a two-wheeled vehicle propelled by a seated rider resting his feet on two pedals and swinging them back and forth. The pedals connected to an articulated rod arrangement, which in turn connected to a large-diameter rear wheel. The rod arrangement converted the back-and-forth pushing of the pedals into rotational motion of the rear wheel. The rider steered the vehicle by turning a handle bar connected to the front wheel. The front wheel measured thirty inches in diameter, and the larger back wheel was forty inches in diameter. Both wheels had iron rims supported by stiff spokes emanating from the centers. In 1842 Macmillan proved that his invention, the first self-propelled bicycle, traveled at least as fast as a post carriage.

Macmillan's vehicle never attracted many buyers. Later, the French father-and-son team Pierre and Ernest Michaux conceived the idea of connecting the pedals directly to the front wheel and thus turning the wheel by rotating the pedals. This design proved popular. In 1862 they built 142 of these "velocipedes." By 1865 they manufactured 400 bicycles per year, which appeared all over Europe. A year later, their mechanic, Pierre Lallement, emigrated to America and took out the first U.S. patent on the bicycle. Improvements on the basic design soon followed. In 1874 H. J. Lawson designed a bicycle whose pedals connected to the rear wheels via an endless chain that engaged sprockets on the pedal drive and the rear wheel. Both front and rear wheels had the same diameter. Much superior than previous designs in terms of stability, braking, and mounting, and called the *safety bicycle,* it set the basic design for almost a century. Finally, in 1888 a Belfast veterinary surgeon, J. B. Dunlop, invented the pneumatic tire, which found its primary early use on the bicycle. By 1890 the bicycle reigned as one of the most popular modes of transportation, infinitely cleaner than the conventional mode at that time—the horse. The advent of the bicycle stimulated new road development, especially in the United States. Bicycles also stimulated interest in lightweight but strong materials and structural designs.

of other bicycle manufacturers. Instead, they handcrafted the individual pieces, paying close attention to detail. This craftsmanship would later carry over to their flying machines.

Thus 1899 found the brothers Wright firmly entrenched in their modest bicycle business and living rather comfortably at 7 Hawthorn Street in Dayton. The Wrights functioned as a team, embodying well-read intellectuality, innate mechanical talents, a penchant for quality craftsmanship, a staunchly moral attitude, and an extremely dedicated work ethic. What fertile soil in which to sow the seeds of the engineering challenge of powered flight.

The Wrights' most important and original contribution to the technology of flight occurred very early in their aeronautical development program—the appreciation of the need for lateral control and a mechanical mechanism for achieving it. Although the Wrights were to pioneer many aspects of aeronautical engineering and applied aerodynamic research, their early recognition of the importance of lateral control and the invention of a mechanical means of achieving it is perhaps the crowning single element of their technology.

Lateral motion of an aircraft is, by definition, the rolling motion about an axis aligned through the fuselage from front to back. Rolling is induced when the lift on one wing is reduced and the lift on the other wing is increased, causing the airplane to roll about the axis. In a modern airplane, the common mechanical device for inducing rolling motion is control surfaces called ailerons. Ailerons are flaplike elements at the trailing edge of the wings; if an aileron is deflected downward, the lift on that wing is increased, and, conversely, if the aileron is deflected upward, the wing lift is decreased. The net imbalance of lift on the two wings results in a rolling motion about the fuselage.

The Wrights clearly knew that Lilienthal achieved lateral and longitudinal control by shifting and swinging his body as it hung beneath his glider. In this way Lilienthal kept the aircraft's center of gravity (including his body weight) and the aerodynamically generated center of pressure (that point on the aircraft through which the net aerodynamic force acts) on top of each other—the necessary condition for equilibrium flight in pitch, roll, and yaw. Wilbur described this difficult problem in a paper he delivered to the Western Society of Engineers in Chicago on September 18, 1901, where he commented on the overall problem of

stability and control: "The balancing of a gliding or flying machine is very simple in theory. It merely consists of causing the center of pressure to coincide with
the center of gravity. But in actual practice there seems to be an almost bound-
less incompatibility of temper which prevents their remaining peaceably together
for a single instant, so that the operator, who in this case acts as peacemaker, often
suffers injury to himself while attempting to bring them together." This clear,
accurate statement of the phenomenon is but one example of the Wrights' basic
understanding of the problems of flight. (Wilbur Wright's writing was clear,
knowledgeable, and to the point. His papers, some showing his subtle humor, still
stand as an excellent example of the art of good technical writing.)

The Wrights rejected the hang glider technique immediately. Adjusting the
center of gravity by swinging one's body seemed too ad hoc, and it was inherently
dangerous due to physical fatigue. Besides, larger and heavier hang gliders had a
limitation; the heavier the machine, the smaller the effect of a person's body weight
(which is a fixed value) on the movement of the overall center of gravity. Therefore,
in Orville's words: "We at once set to work to devise a more efficient means of
maintaining the equilibrium." The Wrights found an intellectual and mechanical
solution to this problem. First came the understanding that a rolling moment,
hence lateral control, could be obtained by simultaneously setting one wing at one
angle of attack to the flow, and the other wing at another angle of attack, such that
the different lift forces on the two wings would induce a rolling motion.

Early in 1900, in a letter to Octave Chanute, Wilbur stated: "My observation
of the flight of buzzards leads me to believe that they regain their lateral balance,
when partly overturned by a gust of wind, by a torsion of the tips of the wings."
This statement has led some people to think that the concept of differentially
changing the angle of attack of the two wings came to Wilbur as a result of observing bird flight. Much later, though, Orville dispelled this idea in a 1941 letter: "I
cannot think of any part bird flight had in the development of human flight
excepting as an inspiration. Although we intently watched birds fly in a hope of
learning something from them I cannot think of anything that was first learned
in that way. After we had thought out certain principles, we then watched the bird
to see whether it used the same principles. In a few cases we did detect the same
thing in the bird's flight. Learning the secret of flight from a bird was a good deal

like learning the secret of magic from a magician. After you once know the trick and know what to look for you see things that you did not notice when you did not know exactly what to look for." So the precise time and manner at which the intellectual solution to lateral control occurred to the Wrights is not known, but it came sometime before July 1899, the month in which they tested the concept using a kite.

Their first mechanical concept for obtaining a differential angle of attack of the two wings involved pivoting both wings via a gear-and-shaft arrangement. One wing pivoted so as to increase the angle of attack and hence increase the lift on that wing, and the other wing simultaneously pivoted in the reverse direction so as to decrease the angle of attack and hence decrease the lift on it. The difference in lift on the two wings set up a rolling motion. They rejected this approach because, as Orville stated much later, "we did not see any method of building this device sufficiently strong and at the same time light enough to enable us to use it."

Wilbur conceived their mechanical solution to obtaining lateral control almost by chance. A customer dropped into the Wrights' bicycle shop to buy an inner tube for a tire. Wilbur removed the inner tube from its cardboard box and began to casually twist the box between his fingers while talking with his customer. He became aware that when he pressed the corners of the long box together in a certain manner, one end of the box flexed downward and the other upward. If the box were an airplane wing, this would be much like the orientation of the wing necessary to obtain an imbalance of the lift over the span of the wing, hence inducing a rolling moment. The history of technology is peppered with instances of serendipitous breakthroughs—this is one such case. The Wrights' invention of twisting the wingtips came to be known as *wing warping*. To achieve this effect, they removed the fore-and-aft diagonal wire bracing of Chanute's trussed biplane configuration but retained the span-wise trussing and twisted the wings across the chord by wires controlled by the pilot. They used wing warping in all of their flying machines.

They proved the viability of wing warping for lateral control simply, rapidly, and, to their satisfaction, less than two months after Wilbur had written to the Smithsonian. They designed a kite made of two superimposed planes, each with a span of five feet and a chord of thirteen inches (giving a wing aspect ratio of 4.62). They controlled the kite by string from two sticks, one held in each hand;

they warped the wings by simultaneously moving the top of one stick forward and the top of the other backward.

Wilbur flew the kite sometime during the last week in July—he did not record the precise date. A group of schoolboys were the only witnesses. Wilbur liked the results; the kite responded to the warping of the wings, always lifting the wing that was twisted upward. Orville said, "We felt that the model [the kite] had demonstrated the efficiency of our system of control." Here we see the first demonstration of the Wrights' engineering methodology, in this case the direct use of a flight test to prove a vital concept. The success of their kite tests encouraged the Wrights to take the next step. They made the decision to construct a machine that was large enough to carry a person but embodied the basic design features of their kite.

In the beginning, Orville and Wilbur did not have large blocks of time available for working on flying machines. As their moderately successful bicycle shop provided their only source of income, their work there had to take precedence over aeronautical work. Their interest and enthusiasm for flying machine experiments grew rapidly, however, and in a letter to Octave Chanute on May 13, 1900, Wilbur wrote: "For some years I have been afflicted with the belief that flight is possible to man. My disease has increased in severity and I feel that it will soon cost me an increased amount of money if not my life. I have been trying to arrange my affairs in such a way that I can devote my entire time for a few months to experiment in this field." The bicycle business regularly peaked in the summer, providing a block of time for the Wright's aeronautical work between September and January. Their intellectual developments continued non-stop; Wilbur's mind in particular constantly dwelled on the technical aspects of flight.

Prior to September 1900 they found time to design a glider that they felt was large enough to carry a person into the air. With the exception of wing warping for lateral control (uniquely their development), they used existing technology. A biplane configuration strengthened by a strut and wire truss system between the wings, the glider design followed that championed by Octave Chanute in the 1890s, with some modifications. The Wrights chose to place a horizontal surface, much smaller than the wing surface, several feet ahead of the wings to help the longitudinal balance of the machine. The location in front of the wings instead of in the more conventional horizontal tail position behind the wings reflected an intentional effort to insure more safety. Always conscious of the gliding accident that killed Lilienthal in 1896, the Wrights reasoned that a horizontal surface placed ahead of the pilot would help to compensate for a rearward shift in the center pressure on the wing caused by unexpected pitch-up due to wind gusts. The forward tail location, called a canard configuration, helped them when the machine would stall by allowing the machine to descend rather gently in an almost "parachute" style. (This is in contrast to the nose dive Lilienthal suffered when his glider stalled, fatally injuring him in the crash.) At the time they designed the 1900 glider, the Wrights had no understanding of the nature of stall and certainly no knowledge of any aerodynamic advantages (or disadvantages) of the canard configuration. (Their later flight experience convinced the Wrights' that the canard configuration worked nicely for them, and they stuck with it for all their flying machine designs until 1910, when in their Wright Model B they placed the horizontal tail in the rear.)

The Wrights used Lilienthal's table of normal and axial force coefficients, from which they easily obtained the lift and drag coefficients to calculate the aerodynamic performance of the 1900 glider. They had access to this table via Octave Chanute's paper "Sailing Flight," printed in 1897 in the *Aeronautical Annual,* in which he republished the Lilienthal table. The original table appeared in 1895 in *Moedebeck's Handbook* in German. The Wrights gained great respect for Lilienthal during their study of the literature, and they felt comfortable in using his aerodynamic data for the design of their flying machine. This general respect for Lilienthal remained with them over the years.

The Wrights designed their 1900 glider to produce enough lift to sustain the machine plus pilot in the air when the airflow velocity relative to the machine was eighteen miles per hour. At this flight velocity (airspeed) and for the estimated weight of the machine, their calculations based on the Lilienthal table indicated that the total wing planform area (the area of both wings combined) should be 200 square feet. These calculations assumed an angle of attack of three degrees—a value chosen by the Wrights as being reasonable. They knew that at higher angles of attack, the drag would be much larger—an undesirable feature. Right from the beginning, the Wrights functioned as reasoned aeronautical engineers—they carried out analysis and made calculations in order to intelligently design their glider. For these early calculations, however, they took off-the-shelf technology and selectively used the pertinent results from the existing literature, such as Lilienthal's table.

On September 6, 1900, Wilbur left Dayton's Union Station for Kitty Hawk, North Carolina, with prefabricated parts of the glider packed in his baggage. He intended to carry out flight tests with the glider at Kitty Hawk, a choice based on U.S. Weather Bureau data on locations where the prevailing winds average over eighteen miles per hour. Left behind to handle bicycle shop business, Orville planned to join Wilbur as soon as he could get the glider ready for flying. The train took Wilbur as far as Old Point, Virginia; from there he took a steamboat to Norfolk. There he tried to find some spruce to make the long wing spars; unsuccessful, he ended up with white pine. Their wing design called for eighteen-foot lengths of spar material, and the white pine had only a maximum length of sixteen feet, forcing Wilbur to reduce the wing area from 200 square feet to 165 square feet. Based on the Lilienthal table, Wilbur calculated that the higher wind velocity necessary for sustaining the redesigned glider at a three-degree angle of attack would be twenty-one miles per hour. From Norfolk he took the train to Elizabeth City, carrying his glider materials with him. On the final leg of the trip, crossing Albemarle Sound to Kitty Hawk, gale force winds and high waves almost sank the unseaworthy boat. Extremely harrowing for Wilbur, what should have been an overnight trip was extended by almost two days from the time Wilbur left Elizabeth City to the moment he knocked on the door of William J. Tate, at whose home he planned to stay during part of that season. Considered to be the most educated person in the small town, Bill Tate served as the unofficial leader of Kitty Hawk. A fisherman, postmaster, notary public, local county commissioner, and member of the local lifesaving crew, Tate welcomed Wilbur, who had survived what was perhaps the most dangerous and life-threatening experience of his life.

Wilbur immediately set about constructing the glider, describing his progress and technical objectives in a September 23 letter to his father:

> I have my machine nearly finished. It is not to have a motor and is not expected to
> fly in any true sense of the word. My idea is merely to experiment and practice with
> a view to solving the problem of equilibrium. I have plans which I hope to find much
> in advance of the methods tried by previous experimenters . . . I am constructing

my machine to sustain about five times my weight and am testing every piece. I
think there is no possible chance of its breaking while in the air . . . My machine will
be trussed like a bridge and will be much stronger than that of Lilienthal, which,
by the way was upset through the failure of a movable tail and not by breakage of
the machine . . . My trip would be no great disappointment if I accomplish practically nothing. I look upon it as a pleasure trip pure and simple, and I know of no
trip from which I could expect greater pleasure at the same cost.

Here is a wonderful indication of Wilbur's innate engineering philosophy. Not
going for the whole prize, he constructed a glider that, for the most part, reflected
the existing nineteenth-century technology but was structurally more sound and
contained one major engineering advancement—wing warping for the purpose
of lateral control and balance. Wilbur wanted to obtain only experience and data.
He looked to make a logical and incremental contribution to the state of the art
of aeronautical technology.

On September 28, Orville arrived at Kitty Hawk. By October 4, they had set
up their own camp in a tent about a half mile away from Tate's house. There they
set about finishing the construction of their glider. Because Wilbur had not been
able to find wood long enough to make a wing spar according to their original
design, they had to be satisfied with wings of smaller area. Each rectangular wing
had a span of sixteen feet and a chord of five feet, giving an aspect ratio of 3.4.
They covered the wings with French sateen fabric and put it on the bias, thus elim-
inating the need for wires to brace the wing surface diagonally. Already sensitive
to the importance of aerodynamic drag reduction, they buried the wing ribs and
the rear spar, which had a square cross section, in pockets of fabric sewn over the
top and bottom of the wood sections in order to locally create a faired-in shape to
reduce the drag. They used fifteen-gauge steel wire for the truss mechanism and
placed wooden vertical supports, called uprights, between the wings.

The Wrights fully understood the consequence of the smaller wing area. They
originally designed the glider to generate sufficient lift to fly at a three-degree angle
of attack in an eighteen-mile-per-hour wind. Now, with the smaller wing area, the
amount of lift necessary to overcome the weight could only be obtained at a higher
angle of attack (hence more drag) and/or at a higher wind speed. They were not

The Wrights first glider, 1900. They designed a biplane configuration with a horizontal surface, much smaller than the wing surface, placed several feet in front of the wings. They flew the glider near Kitty Hawk, North Carolina, in the late summer of 1900. Unfortunately, the glider produced far less lift than the Wrights had calculated, and they accomplished only a few piloted flights. Most of the time they flew the glider as a tethered kite in order to measure its aerodynamic characteristics. Courtesy of the National Air and Space Museum.

too worried, however, because their primary purpose was to obtain technical data and, hopefully, to achieve some successful piloted glides in the process.

The glider made its initial flight during the first week of October; the precise date is not known. Tethered at the end of a rope and rising off the ground in the face of the wind, the glider did not perform satisfactorily with Wilbur on board. They quickly moved to flying the glider as an unmanned kite from the ground. These kite flights served as systematic experiments; the Wrights loaded the glider to different weights using various lengths of chain, and measured the drag (then called *drift* by the Wrights and others). On October 10, they tried flying the glider suspended from a derrick that they had constructed. During these tests, a gust of wind carried away the glider and severely damaged it. In an October 14 letter to Katharine Wright, Orville wrote: "We dragged the pieces back to camp and began to consider getting home. The next morning we had cheered up some and began to think there was some hope of repairing it." The Wrights continued with their flight tests. They abandoned the derrick idea, and for most of the remaining experiments, they flew the glider from the ground like a kite. Beginning on October 18, they started some free flights, the first being simply a free flight of the unmanned glider down a hill. Finally, on October 20, Wilbur climbed aboard. With Orville and Bill Tate holding the wing tips and running down the slope of the Kill Devil Hill (where they had moved for the manned flights), the glider generated enough lift to carry Wilbur into the air. Beginning with a few halting flights of about five to ten seconds duration about one foot above the ground, Wilbur gradually managed to stay in the air for fifteen to twenty seconds, covering 300 to 400 feet over the ground. On October 20, they ceased operations, and on October 23 they started their journey home.

Wilbur had done all the manned flying. He had managed a dozen flights into the air and amassed a total of two minutes of flight time. The Wrights found this short length of flying experience disappointing; they had expected to do much more. On the whole, however, they seemed to be somewhat satisfied with their firsthand efforts to learn about the nature of a flying machine. Later, in 1901 in a paper given to the Western Society of Engineers, Wilbur noted the following about their experience at Kitty Hawk in 1900: "Although the hours and hours of practice we had hoped to obtain finally dwindled down to about two minutes, we were very much pleased with the general results of the trip, for setting out as we did, with almost revolutionary theories on many points, and an entirely untried form of machine, we considered it quite a point to be able to return without having our pet theories completely knocked in the head by the hard logic of experience, and our own brains dashed out in the bargain." The Wrights made some advances in aeronautical technology during their experiments at Kitty Hawk in 1900, the most important being the absolute validation of wing warping as an effective mechanical means of achieving lateral control. Although Wilbur achieved only two minutes of total flying time, he proved that lateral control was essential for successful flight and that the creation of differential lift by presenting the left and right wing tips at different angles of attack to the flow provided a viable mechanism to achieve this lateral control.

This finding proved to be the Wrights' most fundamental contribution to the development of the airplane. The Wrights also made many other observations and collected systematic data from their glider. Taken by itself, each piece of data proved interesting but not momentous. Taken as a whole, however, these observations and measurements began to put the brothers technically head and shoulders ahead of anyone else. Their experience with the 1900 glider gave them something to think about over the following months.

The Wrights made other findings in 1900 that ultimately did not contribute to the development of the airplane. They found the drag created by the body of the human pilot lying prone on the glider to be about one-third that of a pilot in a sitting position. The Wrights were happy with the prone pilot position and stuck with it through their 1903 powered machine, but nobody else adopted the prone position. On another matter, earlier experimenters such as Penaud, Lilienthal, and Langley felt wing dihedral to be very important for lateral stability. The Wrights were not convinced. In their earliest 1900 tests, they found that dihedral made their glider very sensitive to sidewise gusts; they felt this was unsatisfactory and went to straight wings for the remainder of their flights that year. Throughout their subsequent development of the flying machine, the Wrights believed that inherent stability was not all that important to successful flight and felt that they could control their machines much easier without dihedral. This approach worked for them but not for others. It did not produce a meaningful contribution to the early technology of the airplane; indeed, it might be viewed as somewhat retrograde. The rest of the aeronautical community did not accept this philosophy, and most airplanes developed in the twentieth century employed dihedral for lateral stability, as needed. Finally, flying the glider as a kite with Wilbur on board, they measured drag, which curiously decreased as the wind velocity increased. In a letter to Chanute, Wilbur disclosed this finding without comment or explanation. Similar to Langley's data that led to the Langley Power Law, the Wrights' data were on the back side of the power curve, where airplanes rarely fly. The Wrights did not understand this phenomenon, and their drag data in this case made no contribution.

Publication of Research Results

Part of the culture of modern research in science and engineering is the prepa-
ration of a paper discussing research results that contribute new knowledge to
the field and then submission of this paper for possible publication in an archive
or professional journal. After a suitable peer review of the paper, the editors of
the journal decide if it is worthy of publication. The drive to publish research
papers is today stronger than at any time in history, partly because the rapid
dissemination of results is so important to the modern research community and
partly because the reputation (and, hence, advancement and promotion) of the
author is intimately connected with the number and quality of such papers.
In 1900 this drive to publish was not nearly as strong. Granted, peer-reviewed
archive journals existed long before 1900, going back at least as far as 1662, when
the Royal Society of London became the first formal learned society in the field
of science. At the time of the Wright Brothers, though, the "publish-or-perish"
syndrome did not drive most authors; people usually published when and only
when they had something of major significance to say.

Of all the technical observations the Wrights made in 1900, the disagreement
between their measurements and calculations of lift on the glider had the strongest
impact on their future work. Wilbur later wrote about the 1900 glider that "it
appeared sadly deficient in lifting power as compared with the calculated lift of
curved surfaces of its size." Their calculations indicated that, with a pilot on board,
a twenty-one-mile-per-hour wind would lift the glider at an angle of attack of
three degrees. The actual case was much worse; it took a twenty-five-mile-per-hour
wind to lift the glider plus pilot, and the angle of attack was twenty degrees. Since
they used the Lilienthal table for their calculations, they began to have their first
misgivings about the accuracy of Lilienthal's aerodynamic coefficients.

In the final assessment, with the exception of the concrete validation of their
wing warping for lateral control, the Wrights' 1900 glider experiments contributed

only incrementally to the aeronautical state of the art. Nevertheless, the Wrights felt satisfied. When they left Kitty Hawk on October 23, they left their glider behind. They already planned to return the following summer with a bigger, better one. (With the Wrights' permission, Bill Tate salvaged the glider for its materials. Mrs. Tate used the French sateen fabric to make dresses for her two young daughters.)

In July 1901, Wilbur published two short technical articles. He mentioned neither one in his diary, so we do not know exactly when he prepared them. Most likely, he wrote them in the glow of excitement soon after they returned from Kitty Hawk in 1900. We also do not know what motivated Wilbur to write these papers. At that time, the Wrights were not about to discuss in the literature the details of their experiments at Kitty Hawk. In their minds such publication would be premature, to say the least. Instead, Wilbur's two papers dealt with specific aspects of flight technology. "Angle of Incidence" appeared in the July 1901 issue of *The Aeronautical Journal,* published by the Aeronautical Society of Great Britain, by far the most prestigious archive journal in its field at that time. Although Wilbur Wright was totally unknown to the worldwide aeronautical community, that his first technical paper was published in such a respected periodical is not surprising. His relatively short paper (three printed pages) stands as an excellent example of his ability to write clear, understandable, succinct, and interesting technical material.

This paper had the sole purpose of clarifying and standardizing the use of the term *angle of incidence,* which had appeared ambiguously in the previous aeronautical literature. Many authors had referred to the angle between the chord line of a wing and the horizontal as the angle of incidence and made technical calculations of the performance of a flying machine based on this angle. Wilbur dispelled this angle as worthless in calculations. Instead, the angle of incidence defined as the angle between the chord of the wing and the *direction of the airflow relative to the wing* (which may not always be horizontal) is the technically significant quantity. This angle, which today we call the angle of attack, is the germane angle that determines in part the aerodynamic force on the airplane and, hence, its performance. Wilbur began his paper not mincing any words: "If the term 'angle of incidence' so frequently used in aeronautical discussions, could be confined to a single definite meaning, *viz.*, the angle at which aeroplane and wind actually meet, much error and confusion would be averted. But many of the best

Wilbur's Thoughts on Angle of Incidence

Wilbur Wright pointed out a fundamental relation that became an underpinning of modern aeronautical engineering. Every student of aeronautical engineering is familiar with the relation:

$$L = \tfrac{1}{2}\, \rho\, V_\infty^2\, S\, C_L$$

where L is the lift of the airplane, ρ is the ambient density of the air into which the airplane is flying, V_∞ is the flight velocity, S is a reference area (usually taken as the planform area of the wing), and C_L is the lift coefficient. For a given airplane, C_L is primarily a function of angle of attack α; we write this as $C_L(\alpha)$, denoting C_L as a function of α. For small to moderate angles of attack, the function is linear; that is, the increase in C_L with an increase in α is a straight line. For an airplane in steady, level flight, the lift must equal its weight, W, and the above equation becomes:

$$W = \tfrac{1}{2}\, \rho\, V_\infty^2\, S\, C_L\,(\alpha).$$

Solving this equation for $C_L\,(\alpha)$,

$$C_L\,(\alpha) = \frac{2W}{\rho\, S\, V_\infty^2}$$

The last equation demonstrates that for a given airplane of given weight and wing area, flying at a given velocity V_∞ at a given altitude (which determines ρ), the value of C_L is locked in as a specific value, given by the equation. In turn, since C_L is a function of angle of attack, this specific value of C_L corresponds to a specific angle of attack, thus proving that the angle of attack of the airplane at a given altitude is determined by its weight, its wing area, and its flight velocity. The equation tells us, moreover, exactly how α depends on these quantities: the angle of attack varies directly as the weight changes, inversely as the area changes, and inversely as the square of the velocity changes. Wilbur Wright understood this relationship at an early stage of his aeronautical maturing. In his 1901 paper in the *Aeronautical Journal* (London, July 1901, pp. 47–49) he discussed the technical significance of "angle of incidence": "Since the formulation of a principle into a rule often serves to fix it more prominently in the mind, the writer ventures to offer the following rule. The angle of incidence is fixed by area, weight, and speed alone. It varies directly as the weight, and inversely as the area and speed, though not in exact ratio." With this basic relation Wilbur made a substantial contribution to the fundamentals of flight, a fact that has gone unheralded by historians of aviation.

writers use this term loosely and inexactly, with the result that their calculations and explanations of phenomena are thereby often rendered of little value, and students are misled." In the development of the technology of early flight, such basic concepts and logical definitions, no matter how simple they seem today, played a crucial role. In this sense, Wilbur's paper made an important contribution to anyone who was paying attention. The last paragraph of his paper contained a rule stating that the angle of incidence for a flying machine varies in direct proportion to the weight divided by the area and the speed squared. Wilbur was the first to point out this basic relation in aeronautical engineering.

Wilbur's second paper, also published in July 1901, appeared in the German journal, *Illustriete Aeronautische Mitteilungen.* This paper proved to be of only local and transient interest. Only two pages in length, "The Horizontal Position During Gliding Flight" argued for the advantages of a pilot in the prone position rather than in the upright position others used. Even the Wrights abandoned the prone position when they designed and flew their Type A machine in 1908. In light of the overall evolution of the technology of early flight, this second paper by Wilbur had no lasting value.

6

Technology

Further Trials

If you are looking for perfect safety, you will do well to sit on a fence and watch the birds; but if you really wish to learn, you must mount a machine and become acquainted with its tricks by actual trial.

—Wilbur Wright, September 18, 1901, in a presentation to the Western Society of Engineers, Chicago

A NEW GLIDER

In 1901, a busy year for the Wrights, they set about designing a new glider. They desperately needed more lift at lower airspeeds, so they made the glider bigger. It had almost twice the wing area of the 1900 version, 290 square feet rather than 165, and it (with pilot) weighed 240 pounds as opposed to 190. In terms of the wing loading (defined as the weight of the airplane divided by the area of the wing planform), the 1901 glider had a value of 0.83 pounds per square foot, compared to 1.15 pounds per square foot for the 1900 glider.

Rapidly maturing as aeronautical engineers, the Wrights intentionally designed their new glider with a smaller wing loading because they knew that it could therefore lift off the ground at a lower velocity. They incorporated a second feature that they hoped would improve the lift; they increased the airfoil camber to $^1/_{12}$ rather than using the value of $^1/_{23}$ for the 1900 glider. Otto Lilienthal's table applied to an airfoil with a camber of $^1/_{12}$, and the Wrights thought that by matching his value they would remove some of the earlier discrepancy between their calculations and what they actually measured. At the same time they hedged their bets; they designed a mechanism in the wings that would allow them to readily change the camber in the field from one flight to the next, simply by retrussing the wings. With a wingspan of twenty-two feet, the new glider had an aspect ratio of 3.3, smaller than the 3.5 aspect ratio of their earlier glider. The Wrights still did not appreciate the aerodynamic value of high-aspect-ratio wings, and their new design went in the wrong direction in this respect. Otherwise, the new glider had essentially the same features as the 1900 glider.

On May 12, 1901, Wilbur wrote Octave Chanute: "Our plans call for a trip of about six or eight weeks in September and October at the same locality we visited last year on the North Carolina coast." However, the Wrights were so anxious to test their new glider that they left Dayton on July 7, 1901, more than two months earlier than they had originally planned and two months earlier than the previous year. At Kill Devil Hills, about four miles south of Kitty Hawk, they set up camp consisting of a tent and a shed large enough to hold the new glider and to serve as a workshop. They finished the construction of the flying machine on July 27, and that day Wilbur made seventeen glides in the face of prevailing winds of about thirteen miles per hour. Although from the point of view of an independent observer these glides appeared to be successful, one of them covering a distance of 300 feet, Wilbur knew differently. Almost uncontrollable in pitch, on its first few glides the machine nosed almost immediately into the ground. Wilbur, lying prone, shifted his position (hence the center of gravity) further back on the wing for several subsequent flights, until finally he found a position that placed the center of gravity at the center of pressure; the center of pressure was clearly a foot behind where they had expected it to be. With the center of gravity in a more favorable position, the next flight covered over 300 feet, remaining airborne for nineteen seconds, and appeared successful. But Wilbur had to deflect the horizontal canard surface ahead of the wings the full amount, back and forth, to correct for an undulating motion. Unhappy about this, he later wrote: "To the onlookers this flight seemed very successful, but to the operator it was known that the full power of the rudder [the Wrights' term for the forward-placed, movable horizontal canard surface] had been required to keep the machine from either running into the ground or rising so high as to lose all headway. In the 1900 machine one fourth as much rudder action had been sufficient to give much better control. It was apparent that something was radically wrong, though we were for some time unable to locate the trouble."

The Wrights soon diagnosed the problem. They knew that the center of pressure of any airfoil shape changes its location as the angle of attack is changed. For a cambered airfoil, starting at a very high angle of attack (say ninety degrees), the center of pressure is near the middle of the airfoil. As the angle of attack decreases, the center of pressure moves forward, toward the leading edge. At some relatively small angle of attack, the movement of the center of pressure reverses itself, and at yet smaller angles of attack the center of pressure moves rapidly toward the trailing edge as the angle of attack is further decreased. The larger the camber of the airfoil is, the larger the angle of attack at which this reversal occurs, and the faster the travel of the center of pressure. With the increased camber of the 1901 glider wings, the Wrights experienced more movement of the center of pressure, which made their machine more difficult to control.

Wilbur and Orville did not just theorize about this; they also carried out some tests to prove their theory. They removed the upper wing from the glider and flew it as a tethered kite in the prevailing wind. For different wind velocities, the tethered glider took on different angles of attack, with smaller angles of attack corresponding to higher wind velocities. The center of pressure correspondingly shifted as the angle of attack changed. The Wrights were able to use the tethering chord to sense the location of the center of pressure relative to the center of gravity. Wilbur wrote in his 1901 paper to the Western Society of Engineers (with accompanying sketches) that when the center of pressure was ahead of the center of gravity, the chord attached to the leading edge of the wing was pulled upward. When the center of pressure was exactly at the center of gravity, the chord was horizontal. When the center of pressure moved behind the center of gravity, the chord took on a downward orientation. The Wrights used this simple but ingenious test to examine how rapidly and how much the center of pressure was moving on their highly cambered ($^1/12$) wing. They found it to be unacceptably large. Recall that with much prescience they had designed their 1901 glider such that the airfoil camber could be easily changed in the field. They immediately decreased the camber to $^1/17$. This fixed the longitudinal stability problem, and for the next glides the machine handled more like their 1900 glider, much to the relief of the Wrights.

The Wrights measured the local wind speed with a small, handheld anemometer manufactured by Richards in Paris; this device had small, flat vanes that rotated in the wind. Chanute had recommended the Richards and loaned his to the Wrights. They attempted to measure the accuracy of and calibrate their anemometer by comparison with other measurements. The Kitty Hawk Weather Station had an anemometer mounted thirty feet above the ground for measuring wind speeds. They found that the Richards anemometer registered wind speeds that were too high by about 20 percent, and they took this calibration into account when recording wind speeds. (When they constructed the 1903 flyer, they mounted a Richards hand anemometer on the front center strut adjacent to the pilot's body, for the purpose of measuring the airspeed.)

Much to their consternation, the Wrights found that the lift of their new glider was as woefully deficient as that of the 1900 model. Although the 1901 glider had a larger wing area and produced more lift than their smaller 1900 one, the measured lift still fell far below their predictions based on Lilienthal's table. Wilbur wrote in his diary on July 29, 1901: "Afternoon spent in kite tests. Found lift of machine much less than Lilienthal tables would indicate, reaching only about $^1/3$ as much." The next day, he wrote: "The most discouraging features of our experiments so far are these: The lift is not much over $^1/3$ that indicated by the Lilienthal tables. As we had expected to devote a major portion of our time to experimenting in an 18-mile wind without much motion of the machine, we find that our hopes of obtaining actual practice in the air are decreased to about one fifth of what we hoped, as now it is necessary to glide in order to get a sustaining speed. Five minutes' practice in free flight is a good day's record. We have not yet reached so good an average as this even." Wilbur went on to list other discouraging features such as poorer control compared to their 1900 glider, a higher measured drag (almost double) than expected, and more sluggishness in free flight.

The much smaller lift ruined any hope of long hours of flight testing with the glider tethered as a kite and a pilot aboard. The Wrights had counted on this mode of operation to give them long hours of practice controlling the machine in the air. It was not to be. Although Wilbur achieved far more free flight gliding time in 1901 than he had the year before, by no means did he feel satisfied with the state of affairs.

During part of their stay at Kill Devil Hills, they had visitors. Octave Chanute arrived on August 5 and remained for six days, taking close notes of the experiments, snapping photographs of the machine in flight, and generally providing good fatherly company for Wilbur and Orville. Two other visitors, Edward Huffaker and George Spratt, joined them at Chanute's suggestion. Huffaker had experience with flying machines, having worked as an assistant to Samuel Langley at the Smithsonian between 1894 and 1898. Spratt was a Pennsylvania farmer with an informed interest in flying machines. The Wrights soon found Huffaker disagreeable, but they liked Spratt, with whom they remained close friends, and he and Wilbur frequently exchanged letters over the ensuing years.

Their spirits dampened by rain and their growing despair about their test results, the Wrights left for home on August 20. Wilbur later recalled that on the trip home he dejectedly said to Orville "that men would not fly for fifty years." In 1940 Orville remembered this statement as "not within a thousand years would man ever fly." Back in Dayton, they struggled to make sense of their flying experience. Fortunately, Chanute provided an almost therapeutic opportunity for the Wrights. Impressed by what he saw during his visit to Kill Devil Hills, on August 29 Chanute wrote to Wilbur: "I have been talking with some members of the Western Society of Engineers. The conclusion is that the members would be very glad to have an address, or a lecture from you, on your gliding experiments. We have a meeting on the 18th of September, and can set that for your talk. If you conclude to come I hope you will do me the favor of stopping at my house." (Note that Chanute made the invitation to Wilbur, whom he obviously recognized as the intellectual leader of the Wrights' team effort. Chanute directed virtually all of his correspondence to Wilbur. Also, Wilbur had piloted the 1900 and 1901 gliders; Orville did not get into the air until the following year.)

Wilbur at first hesitated to accept Chanute's invitation. Their flight experiments to date introduced so many unanswered questions, and Wilbur had never given a technical paper to a professional audience. Nevertheless, the receipt of such a kind invitation from a man of Chanute's stature uplifted both brothers at a time when they felt quite discouraged about their perceived lack of progress. On September 2 Wilbur replied with a mild *yes* to Chanute's invitation. Their sister Katherine, always a staunch supporter and close companion to Wilbur and Orville, played a strong role in his decision. In a September 3 letter to her father, she wrote: "Will was about to refuse but I nagged him into going. He will get acquainted with some scientific men and it may do him a lot of good." (A definitive study of Katharine's behind-the-scenes role in the success of the Wright brothers does not exist and would make a great research project.)

Although he had only a few weeks to prepare his paper, this period proved an excellent opportunity for Wilbur to collect his thoughts and reflect on the results of their flight experiments. At 8 P.M. on September 18, 1901, Wilbur stood before a friendly and willing audience of the Western Society of Engineers in Chicago and delivered a profound technical paper. Simply entitled "Some Aeronautical Experiments," the paper represented a comprehensive survey and commentary on their progress to date. Wilbur ended by itemizing what seemed to him their most important technical findings:

In looking over our experiments of the past two years, with models and full-sized machines, the following points stand out with clearness:

1. That the lifting power of a large machine, held stationary in a wind at a small distance from the earth, is much less than the Lilienthal table and our own laboratory experiments would lead us to expect. When the machine is moved through the air, as in gliding, the discrepancy seems much less marked.
2. That the ratio of drift to lift in well-shaped surfaces is less at angles of incidence of five degrees to 12 degrees than at an angle of three degrees. [The ratio of drift to lift is the reciprocal of the lift to drag ratio, L/D. Wilbur is noting that the lift to drag ratio is higher at angles of attack between five and twelve degrees than it is at lower angles of attack. Using Wilbur's estimate that the maximum L/D for the 1901 glider with pilot aboard was about seven, my analysis shows that the angle of attack for maximum L/D is about eight degrees. This data point is con-

sistent with Wilbur's observation; as usually was the case, he knew what he was talking about.]

3. That in arched surfaces the center of pressure at 90 degrees is near the center of the surface, but moves slowly forward as the angle becomes less, till a critical angle varying with the shape and depth of the curve is reached, after which it moves rapidly toward the rear till the angle of no lift is found. [Here, Wilbur is spelling out for the first time in the literature what later became a well-known characteristic of cambered airfoils.]

4. That with similar conditions, large surfaces may be controlled with not much greater difficulty than small ones, if the control is effected by manipulation of the surfaces themselves, rather than by a movement of the body of the operator. [With the few successful flights of their 1901 glider, Wilbur had flown the largest and heaviest flying machine in history. He is pointing out that a large machine can be safely flown, but with mechanical controls rather than by shifting body weight as Lilienthal had done.]

5. That the head resistances of the framing can be brought to a point much below that usually estimated as necessary. [Head resistance, a term the Wrights adopted from the current aeronautical literature, later became identified as parasite drag, which can be reduced by proper streamlining of the airplane shape. Wilbur's statement reflects two points. First, they reduced the drag of the 1901 glider by locating the leading edge wing spar underneath the curved airfoil nose so as to present a more streamlined shape to the flow, making their glider more aerodynamically efficient. Second, the existing calculational techniques for head resistance, being somewhat ad hoc, fell subject to considerable error. Wilbur is simply saying two things: that their attention to drag reduction paid some dividends, and that the existing techniques for calculating head resistance resulted in values that were too large.]

6. That tails, both vertical and horizontal, may with safety be eliminated in gliding and other flying experiments. [Here, Wilbur is touting their design with the forward-placed horizontal canard surface. They had no rearward-placed tail in the more conventional sense as used by George Cayley, Alphonse Penaud, Lilienthal, Chanute, and Langley. Wilbur is simply saying that their configuration worked. The Wrights clung to their canard design until their Model B in 1910, when they moved the horizontal surface to the rear of the airplane. In 1902, however, they

found it necessary to mount a vertical tail surface in the rear for all of their subsequent machines.]

7. That a horizontal position of the operator's body may be assumed without excessive danger, and thus the head resistance reduced to about one fifth that of the upright position. [This is simply a reiteration of the point made by Wilbur in his earlier paper published in the *Illustrierte Aeronautische Mitteilungen*.]

8. That a pair of superposed, or tandem surfaces, has less lift in proportion to drift than either surface separately, even after making allowance for weight and head resistance of the connections. [This is the first statement in the aeronautical literature of the adverse interference effect due to the biplane configuration. The Wrights already recognized that by placing one wing above the other or one directly behind the other, the lift of the combined wings is less and the drag is more in comparison to a single wing with the same total area and the same aspect ratio. The biplane interference effect became a subject for intensive theoretical study in the period around 1918–20. Amazingly, the Wrights discovered this effect in 1901.]

The lack of agreement between their measurements of lift and their calculations based on the Lilienthal tables perplexed the Wrights. In spite of their obvious respect for Lilienthal, their suspicion of his tables became so strong that Wilbur boldly wrote in his paper to the Western Society that "the Lilienthal tables might themselves be somewhat in error." That fall, the Wrights became even more convinced. (The question as to the validity of Lilienthal's tables continued to recent times. Recent research, however, explains and reconciles the discrepancy.[1] Simply stated, the Wrights misinterpreted the tables in three respects, and when these misinterpretations are accounted for, the tables predict a lift about one-third that calculated by the Wrights—exactly what the Wrights measured. Because they help to put the technology of early flight in perspective, details of these new findings are summarized in Appendix B.)

WORKING IN THE LABORATORY

The discrepancies between their calculations and measurements exacerbated Wilbur and Orville's depressed state as they returned home from Kitty Hawk in

August 1901, but their discontent soon faded. Before Wilbur gave his paper to the
Western Society, the brothers made an intellectually courageous decision. They
had used the best aerodynamic data then available (at least to the extent that they
understood it), and yet their gliders performed considerably below par (in comparison with their calculations). They concluded that something must be wrong
with the existing data. Therefore, they determined to start over and compile their
own aerodynamic data. That decision—essentially to do everything themselves,
starting with the very basics—expedited the success of the Wright Flyer in 1903.

Whether or not the Lilienthal table was accurate, the Wrights *thought* that it
might not be accurate. Wilbur's July 27, 1901, diary entry contained the first written indication of their suspicions. He referred to discrepancies in their drag calculations compared to their measurements, "thus leading to doubts of the correctness of Lilienthal table of ratio of lift to drift." Later, in his paper to the
Western Society of Engineers, he considered "that the Lilienthal tables might
themselves be somewhat in error." This concern strongly motivated them to carry
out their own aerodynamic measurements, as did uncertainty about the value of
Smeaton's coefficient, which Wilbur also questioned in his paper to the Western
Society. In a letter to Chanute on September 26, 1901, Wilbur noted "that Prof.
Langley and also the Weather Bureau officials found that the correct coefficient
of pressure was only about 0.0032 instead of Smeaton's 0.005." These questions
served as a catalyst for the brothers' experiments in aerodynamics. During these
experiments the Wrights found "the right aerodynamics."

Verifying Lilienthal's results was their initial concern. The Wrights first
attempted to make some comparison measurements by mounting a small model
wing with a cambered airfoil on the rim of a horizontally positioned bicycle wheel,
and a flat plate oriented perpendicular to the wind direction at a location on the
rim ninety degrees from the wing. With a wind blowing over the wheel, when the
lift on the wing equaled the drag on the plate, the wheel was balanced; that is, it
did not rotate. The Wrights first tested this apparatus in the natural wind with no
conclusive results. They then mounted the wheel rim horizontally on the front
of a bicycle and peddled the apparatus through the air. They determined from the
Lilienthal table that the curved surface would have to be at a five degree angle of
attack to balance the drag on the flat surface. Their measurements, however,

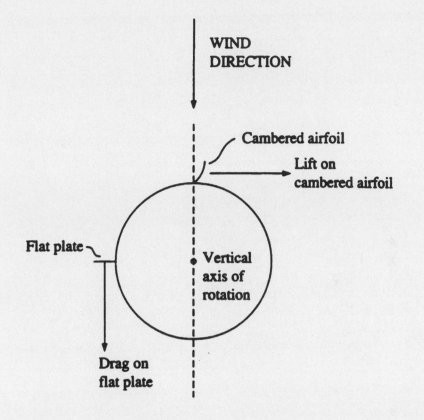

WIND DIRECTION

Cambered airfoil

Lift on cambered airfoil

Flat plate

Vertical axis of rotation

Drag on flat plate

Top-view schematic of the Wrights' setup for the measurement of lift on a wing model attached to the rim of a bicycle wheel. With the wheel oriented horizontally in a wind, when the lift on the airfoil equaled the drag on a flat plate oriented perpendicular to the flow, the wheel remained balanced in the position shown. If the lift and drag were not equal, the wheel would rotate. Author's collection.

showed that balance required eighteen degrees angle of attack—indicating the lift on the curved surface to be less than half that predicted by their calculations using the Lilienthal table. This finding supported their glider tests, where they found the measured lift to be considerably smaller than what they had calculated.

Wilbur recorded some watershed decisions in an October 6 letter to Chanute. He wrote that the wheel rim results prompted them to examine carefully the

Impact of Misinterpretation on Wheel Rim Tests

For their glider calculations the Wrights misinterpreted how to use the numbers from the Lilienthal table; the same problem prevailed for the tests using
the wheel rim. The curved wing surface on the wheel rim had an aspect ratio
of 2.25—far smaller than the aspect ratio of 6.48 used by Lilienthal. Again the
Wrights showed no awareness of the effect of aspect ratio on the lift coefficient.
They made the shape of the airfoil section the same circular arc with $1/12$ camber used by Lilienthal, causing no discrepancy due to a different airfoil shape.
Neither did the wrong value of Smeaton's coefficient play a role here. The
Wrights set up the wheel rim test so that the lift on the cambered wing balanced
the drag on the flat plate, and since the Wrights calculated the lift and drag
using Smeaton's coefficient, that coefficient cancelled out of the balance equation. For their wheel rim tests, the different aspect ratios caused the discrepancy between what the Wrights expected based on the Lilienthal table and what
they actually observed.

sources of error in their earlier glider tests at Kitty Hawk. They become absolutely
convinced that one source of the error was "in the use of the Smeaton coefficient
of 0.005." Wilbur then adopted the value measured by Langley, stating: "I see no
good reason for using a greater coefficient than 0.0033." From that time on, they
used this reasonably accurate value for Smeaton's coefficient. Wilbur also
announced their venture into wind tunnel testing. He described a small, makeshift
wind tunnel that Orville constructed from an old starch box and a small fan turning at 4,000 revolutions per minute; the whole apparatus was eighteen inches long.
They used this small tunnel for only one day, but it proved to be an important
stepping-stone to their next generation of aerodynamic testing. With this apparatus, the Wrights confirmed "the correctness of Lilienthal's claim that curved surfaces lift at negative angles." Again, however, the Wrights' quantitative predictions
based on the Lilienthal table did not agree with their experimental results. With
these makeshift wind tunnel tests in mind, Wilbur wrote: "I am now absolutely

certain that Lilienthal's table is very seriously in error, but that the error is not so great as I had previously estimated." Wilbur informed Chanute of the reasons for their decision to build a much more sophisticated wind tunnel: "The results obtained, with the rough apparatus used, were so interesting in their nature and gave evidence of such possibility of exactness in measuring the value of [lift coefficient], that we decided to construct an apparatus for making tables giving the value of [lift coefficient] at all angles up to 30° and for surfaces of different curvatures and different relative lengths and breadths." With this, the Wrights took a dramatic step. They planned to construct their *own* detailed tables of aerodynamic coefficients; moreover, this would be a whole series of tables for different airfoil shapes and wing planform shapes, going well beyond the Lilienthal table.

Wilbur's letter to Chanute reflected a degree of technical maturity and self-confidence well beyond that of any of their previous correspondence. The brothers now used with conviction a reasonably correct value of Smeaton's coefficient. They felt convinced about the viability of the wind tunnel for producing accurate, detailed data. They stood ready to employ such a tunnel to produce tables of coefficients that would be their own original contribution to the field of applied aerodynamics. The data from these wind tunnel experiments would be as accurate as they could make it—data in which they could have confidence. Finally, these data would be directly applicable to their future glider design. In this regard, the Wrights acted as engineers, focusing on obtaining only that data necessary to accomplish their aim. They felt excited about the prospects for their wind tunnel tests. What a dramatic contrast to the depths of despair they felt just a month and a half earlier.

The new wind tunnel was built and operating by mid-October. As usual, the Wrights did everything themselves, designing and building the wind tunnel out of wood. Much larger than their earlier starch box, the flow duct measured six feet long with a square cross-section sixteen inches on each side. They placed a glass window on top for observing the tests. A fan drove the airflow, powered by the central power plant of the bicycle shop—a one horsepower gasoline engine connected to the fan via shafts and belt drives—creating a maximum velocity in the wind tunnel of thirty miles per hour. The tunnel rested on the second floor of the shop, where all the testing took place. With this tunnel, the Wrights followed in

Replica, based on written descriptions, of the Wright brothers' wind tunnel. During 1901–02, they obtained data from wind tunnel tests on numerous different wing and airfoil shapes. These data proved instrumental in the successful design of their 1902 glider, and subsequently the 1903 Flyer. The Wrights operated the tunnel on the second floor of their bicycle shop in Dayton. They later scrapped the tunnel, and no photograph of it has ever been found. Courtesy of the National Air and Space Museum.

the tradition of Francis Wenham, Horatio Phillips, and a dozen other developers who previously had built and operated wind tunnels. The Wrights' wind tunnel, however, became the second (Hiram Maxim's was the first) used specifically to produce data for the single purpose of designing a flying machine.

They designed two different force balances for use in their wind tunnel, both ingenious and unique. One balance measured the lift coefficient directly (rather than the lift force itself). The mechanical design ensured that uncertainties about flow velocity in the tunnel and the value of Smeaton's coefficient did not affect the final result for lift coefficient. The second balance, a simple design, measured directly the drag-to-lift ratio. Knowing the lift coefficient (from the first balance) and the ratio of drag to lift (from the second balance), the Wrights directly calculated the drag coefficient.

From mid-October to December 7, 1901, the Wrights tested over 200 different wing models, with different planform and airfoil shapes. They tested camber ratios from 1/6 to 1/20; the location of maximum camber ranged from near the leading edge to the midchord position. The planform shapes included squares, rectangles, ellipses, surfaces with raked tips, and circular arc segments for leading and trailing edges meeting at sharp points at the tip. They also examined tandem wing configurations (after Langley's aerodromes), biplanes, and triplanes. Finally, Wilbur and Orville had to end these experiments because of the press of business. In a December 7, 1901, letter to Bishop Wright, Katharine Wright wrote: "The boys have finished their tables of the action of the wind on various surfaces, or rather they have finished their experiments. As soon as the results are put in tables, they will begin work for next season's bicycles." These experiments, conducted over less than a two-month period, produced the most definitive and practical aerodynamic data on wings and airfoils obtained to that date. They gave the Wrights the necessary aerodynamic information to design a proper flying machine.

The Wrights chose to tabulate data from forty-eight of their wing models. One table gave lift coefficient versus angle of attack, α. Another table gave the drag-to-lift ratio as a function of angle of attack. These tables supplanted the Lilienthal table in all respects. At the time, the Wrights' tables represented the most valuable technical data in the history of applied aerodynamics.

TO PUBLISH OR NOT TO PUBLISH

As it turned out, during the seminal period of the birth of the airplane, when the Wrights' tables could have found widespread use, they remained for the Wrights and Chanute's eyes only. Wilbur and Orville originally did not intend to keep their data secret. They fed Chanute bits and pieces of their lift-and-drag measurements during the course of their wind tunnel experiments. In a November

27, 1901, letter to Wilbur, Chanute mentioned that Major Hermann Moedebeck from Germany had invited him to write a new chapter for an updated version of the Moedebeck *Handbook for Aeronauts and Aviators*—the source of the original publication of the Lilienthal table in Berlin in 1895. Chanute gently hinted that some of the Wrights' data might be included in his contribution to the handbook: "Now, I will either prepare the whole of the notes, including your experiments, or prepare the notes up to the latter point, and let you describe your own work, as you may prefer. Please let me know." Wilbur penned his modest reply on December 1: "I think very well of your plan to republish the Lilienthal section of Moedebeck's handbook substantially in its present form and add your own notes as a supplementary article . . . You will be better able than we to preserve a proper perspective in describing our experiments, so you had better keep the matter in your own hands. It is a question whether any table additional to his should be inserted, but if deemed advisable it should preferably be of a surface of a markedly different character so that instead of contradicting Lilienthal it should empha-size the necessity of considering shape, relative dimensions, and profile in calcu-lating the expected performance of a machine." Chanute wrote on December 11: "I quite agree with you that it will be preferable to give in Moedebeck's handbook the coefficients for a surface differing markedly from Lilienthal's, and to empha-size the necessity of considering shape. Please furnish me the necessary data and comments when you consider that you have arrived at such definite results as to warrant publication."

Wilbur did not reply until January 5, 1902, when he sent Chanute tables of aerodynamic data for seventeen different models. In regard to these data and the publication thereof, Wilbur wrote:

In a recent letter you inquired what of our tables I thought ought to be given to Moedebeck. My failure to answer sooner was for the simple reason that I did not know what to say. On the one hand, the value of our tables lies chiefly in the oppor-tunity they afford of comparisons of the effect of aspect, curvature, thickening, and chord, upon the lift, the tangential, and the angle at which the maximum occurs. But on the other, there is the objection that to include all would make more than would be advisable in so brief a work as a handbook. And then there is the further and greater objection that to insert in an authoritative work like a handbook a set of tables which are not claimed to be perfect, in advance of their general public acceptance, would entail a personal responsibility on your part which ought not to be assumed lightly. Although I have great confidence myself in their substantial accuracy, yet there comes the haunting thought that all previous experimenters in this line have made mistakes and that though we have avoided or corrected ninety-nine sources of error there may be one that has escaped attention. We could assume the responsibility of issuing them with a clear conscience, but the case would be somewhat different with you even though you should disclaim personal responsibility. However, when you have figured and studied all our measurements, you will be better able to determine the best course, whether to make a general statement of the tendency of the results, or whether to say nothing at all. We will send on the data of the measurements of the other surfaces shortly.

This reply showed various degrees of modesty, coyness, conservativeness, and deference to Chanute. In a roundabout way, Wilbur still left up to Chanute the decision to publish the Wrights' data, but he gave the impression that he was dragging his feet. Chanute shared Wilbur's reluctance in a February 6 letter, complaining that Moedebeck was rushing him for the contribution to the new handbook. "I had come to quite the same conclusion as yourself," Chanute went on, "i.e., that it would be unwise to give the public full information as to the properties of curved surfaces at present, and I wrote him (Moedebeck) that the article would not contain your data." After this, Wilbur considered the publishing of the tables in the Moedebeck handbook to be a dead matter. Instead, he commented in a February 7 letter to Chanute: "In considering the matter of publishing our tables of pressures and tangentials, . . . I think I shall prepare to make them public sometime during this summer." He never did. After all the correspondence and discussion, the complete aerodynamic tables compiled by the Wrights were never published in their lifetime. (Finally, from the Wrights' notebooks M. W. McFarland compiled the tables for forty-eight different wing models and published them as an appendix in *The Papers of Wilbur and Orville Wright* [1953].)

Keeping in mind that Leonardo da Vinci's notes were not published for centuries after his death, and hence, by default, da Vinci made no contribution to the state of the art in aeronautics during the period in which he lived, we might be tempted to make the same conclusion about the Wrights' aerodynamic data. However, Da Vinci's thinking never resulted in a successful flying machine. The Wrights' did. So we can easily argue that Wilbur and Orville's successful airplanes embodied their aerodynamic data and hence transferred it to the public in a dramatic and compelling fashion—much more dramatic than the publication of their tables would have been.

By December 1901 the Wrights had uncovered what was for them the right aerodynamics. In particular, they had demonstrated to themselves the advantage of high aspect ratio. Their wind tunnel models ranged in aspect ratio from one to ten. Although Langley's results, published in *Experiments in Aerodynamics,* clearly showed that higher aspect ratio wings produced more lift, the Wrights appeared not to have noted Langley's data, or they simply ignored it. The correspondence between Wilbur and Chanute made no mention whatsoever of Langley's aspect ratio data. Their own wind tunnel data demonstrated to the Wrights the advantage of higher aspect ratio wings. Wilbur noted that the model with "the highest dynamic efficiency of all the surfaces shown," had a camber of 1/20 and an aspect ratio of six. The data from this model had a strong effect on the wing design of their 1902 glider.

By the end of 1901 the Wrights possessed far more aerodynamic data and understanding than anyone else before them. In a deposition dated February 2, 1921, Orville stated: "Cambered surfaces were used prior to our experiments. However, the earlier experimenters had so little accurate knowledge concerning the properties of cambered surfaces that they used cambered surfaces of great inefficiency, and the tables of air pressures which they possessed concerning cambered surfaces were so erroneous as to entirely mislead them. They did not even know that the center of pressure traveled backward on cambered surfaces at small angles of incidence, but assumed that it traveled forward. I believe we possessed in 1902 more data on cambered surfaces, a hundred times over, than all of our predecessors put together." In this statement, Orville unintentionally shortchanged Langley, who measured the reversal in center of pressure before the Wrights. Langley, however, never published the data, and the Wrights therefore never knew about it. Orville also did not pay enough respect to the reasonable accuracy of the

Lilienthal tables. In the main, however, he was justified when he emphasized that he and Wilbur were the sole owners of the most advanced, most precise, and most viable data in applied aerodynamics at that time—another reason why the Wright brothers were the first true aeronautical engineers.

In October 1901, Chanute gave the Wright brothers a partially translated version of Lilienthal's *Birdflight as the Basis of Aviation.* The Wrights finally became aware of the shape and aspect ratio of Lilienthal's test model wings and no longer mistakenly considered the Lilienthal table as a set of universal values but properly viewed it as a tabulation of aerodynamic coefficients pertaining only to a certain wing and airfoil shape. Indeed, the Wrights felt they had been a little misled by Chanute, who had republished Lilienthal's table in his paper "Sailing Flight" that appeared in *The Aeronautical Annual* in 1897. Chanute never qualified the limitations on the table, giving the impression that it was universal in nature. In a November 22, 1901, letter, Wilbur gently took Chanute to task: "In the Lilienthal table I would suggest that it be stated in connection with the table that it is for a surface 0.4 meters x 1.8 meters, of the shape [here Wilbur drew the planform shape of Lilienthal's wing model], and a curvature of 1 in twelve, and measured in the natural wind. Any table is liable to great misconstruction if the surface to which it is applicable is not clearly specified. No table is of universal application."

From this time on the Wrights considered the Lilienthal table moot. From their wind tunnel tests, they had their own tables of aerodynamic coefficients, which were much more extensive and detailed than any from the past. Besides, they had a reasonable degree of confidence in the accuracy of their own measurements. They never again used the Lilienthal table for their design calculations.

7

Technology

A Working Aircraft

One of the most gratifying features of the trials was the fact that all our calculations were shown to have worked out with absolute exactness so far as we can see, though we have not yet made our final computations on the performance of the machine.

—Wilbur Wright, December 28, 1903, in a letter to Octave Chanute eleven days after the successful flights of the Wright Flyer

FIELD TESTING AT KITTY HAWK

The Wrights brought their wind tunnel tests to a conclusion by early December 1901 and returned to the bicycle business. During their work on "next season's bicycles," the Wrights found time to design "next season's" glider. Based on knowledge gained from their wind tunnel tests, the new design had wings with an aspect ratio of 6.7—more than twice that for their 1901 glider. The Wrights had learned their lesson well. The new wings had an area not much greater—305 square feet compared to the 1901 glider wing area of 290 square feet. They simply made each wing longer and narrower. They also designed the airfoil camber to be quite small—$^{1}/_{25}$ compared to the value of $^{1}/_{12}$ they had started with in their 1901 flight trials. Even their forward canard surface had an aspect ratio larger than that of the previous year.

The new design had another striking difference—a vertical tail at the rear of the machine, consisting of two side-by-side vertical surfaces, giving a total vertical tail area of 11.7 square feet. They intended the vertical tail to counteract a disturbing yawing motion that Wilbur had sensed in a few of his 1901 glides but had not mentioned in his paper to the Western Society, presumably because he had no plausible explanation for its occurrence. He briefly mentioned the problem in a August 22, 1901, letter to Chanute: "The last week [of flights in 1901] was without very great results though we proved that our machine does not turn [i.e., circle] toward the lowest wing under all circumstances, a very unlooked for result and one which completely upsets our theories as to the causes which produce the turning to the right or left."

If you are piloting an airplane and you wish to turn to the right, you induce a roll to the right (you deflect the ailerons so the right wing drops and the left wing raises up). This rotates the aerodynamic lift vector toward the right, and hence the airplane begins to make a circle to the right. That is, you execute a turn by "pointing" the lift vector toward the direction in which you want to turn. Wilbur understood this principle—this constituted his "theory as to the causes which produce the turning to the right or left." There is a side effect, however, to this rolling motion of the wing. To get the right wing to drop and the left wing to rise, the lift is decreased on the right wing and increased on the left wing—the Wrights accomplished this by wing warping. There is an increase in wing drag due to the production of lift—called *induced drag* or *drag due to lift*. When the lift is increased on the left wing, the drag on the left wing also increases. Similarly, when the lift is decreased on the right wing, the drag on the right wing decreases. This difference in drag causes the airplane to yaw (pivot) towards the left. That is, the nose of the airplane wants to swing toward the left, in the opposite direction of the desired turning direction. This yawing motion is precisely what Wilbur had experienced but had not expected. The brothers did not understand what made it happen, but they correctly theorized that this undesirable yawing action could be suppressed by adding a vertical tail at the rear of the machine. A vertical tail provides a "weather-vane" effect, which would swing the glider back to its proper zero-yaw position if any unexpected yaw motion occurred during a turn. Hence, their new design had a vertical tail (indeed, to begin with, a double vertical tail).

In August, the Wrights worked on the prefabricated parts of their new 1902 glider. On August 20, Katharine wrote to her father that: "The flying machine is in process of making now. Will spins the sewing machine around by the hour while Orv squats around marking the places to sew. There is no place in the house to live but I'll be lonesome enough by this time next week and wish that I could have some of their racket around." After arriving at their camp at Kill Devil Hills on the twenty-eighth, the brothers set about improving their living conditions. They found their shed from the previous year in disrepair, so they remodeled and enlarged it, as well as improved the interior. They dug a sixteen-foot well nearby; Wilbur claimed that: "It is the best [water] in Kitty Hawk." In much better spirits than the previous year, Wilbur wrote to George Spratt on September 16 that "we are having a splendid time."

On September 8 they started to assemble their glider, finishing it eleven days later. After a midday dinner they flew the glider as a kite and found themselves very encouraged by its behavior. Orville noted in his September 19 diary entry that: "We made no entirely free flights, but from several glides made are convinced that the machine will glide on an angle of seven degrees, or maybe less." Over the next two days, this proved to be the case. Wilbur made nearly fifty glides, most of them at angles between 7 and 7.5 degrees. This is to be compared with glide angles of around ten degrees or higher for their earlier 1901 glider. The angle of glide (the angle between the descending flight path of the glider and the horizontal) is strictly a function of the lift-to-drag ratio of the glider. Trigonometrically, the tangent of the glide angle is equal to the inverse of the lift-to-drag ratio. The higher the lift-to-drag ratio is, the more shallow the glide angle.

Wilbur knew this principle. In his 1901 paper to the Western Society, he succinctly stated: "In gliding experiments, however, the amount of lift is of less relative importance than the *ratio* of lift to drift (drag), as this alone decides the angle of gliding descent." Their 1902 machine, gliding at the smaller angle of about seven degrees, clearly had a much improved lift-to-drag ratio, primarily due to its much larger wing aspect ratio.

On September 23, Orville made his first free-flight glide. Prior to that, Wilbur had made all the free gliding flights for all their machines. Now the Wrights shared the piloting of the machine. They set their primary goal to fly as long and as frequently as possible in order to gain experience in the air. During the almost two months that they spent at Kill Devil Hills that year, they finally accomplished their goals for gliding, successfully executing somewhere between 700 and 1,000 flights, the longest being twenty-six seconds in the air covering 623 feet over the ground.

In terms of advancing the technology of early flight, the Wrights' 1902 glider made three important contributions: (1) the clear demonstration in flight of the aerodynamic efficiency of high-aspect-ratio wings, (2) the use of the vertical tail, and (3) the structural design of the glider. We have already discussed at length the effect of the higher aspect ratio. The other two contributions deserve more discussion.

Fixed vertical tails were not new—the concept went back as far as George Cayley's work in 1799. During the course of their flying in 1902, the Wrights made a modification to the vertical tail that was pivotal (no pun intended); they modified the tail so that it could be rotated about its vertical axis, thus acting as a moveable rudder. The Wrights' intelligence in dealing with the circumstances that led to this change is yet another example of their logical engineering talent. The 1902 glider started with a fixed double vertical tail that the Wrights correctly reasoned would solve the adverse yaw problem. As they began their flights, they found that this particular problem had indeed been solved. In fact, the Wrights shortly removed one of the vertical surfaces because a single vertical tail proved just as effective. Yet now a new problem surfaced. Whenever an airplane rolls (banks), it also sideslips (makes a sideways motion) in the direction of the lower wing. Therefore, the entire side of the airplane that faces into the sideslip feels a portion of the relative air velocity directed at right angles to the side. This includes the fixed vertical tail, which feels force perpendicular to its surface due to the sideslip. In turn, this swings the tail away from, and hence the nose into, the direction of the sideslip. If left uncontrolled, this can set the airplane spiraling into the turn, that is, making a corkscrewing motion around the tip of the lower wing—a spin. Spins happened a few times during the early flights in 1902, with the result that the glider fell out of the sky and the wing tip slammed into the sand. The Wrights called such an event *well digging*.

The manner in which the Wrights solved this problem reveals an intellectual molding together of their two minds. Until this point, Wilbur had been the intellectual drive behind most of their developments and advancements, whereas Orville provided organizational skill as well as served as a sounding board in their many spirited discussions about flying machines. The two of them made a great team. This time Orville had a great idea. Not able to sleep the night of October 3 and mulling over the well-digging problem in his mind, he came to an inspired and straightforward solution to the problem. The next morning at breakfast he shared with Wilbur his idea that the vertical tail should be made to pivot as a rudder. In this way, when the tail started to swing in an undesirable direction in a sideslip, rudder deflection controlled by the pilot would counteract and stop the undesirable yawing motion.

Orville expected Wilbur to say something like "Oh yes, I was already considering that"; as the older brother, Wilbur sometimes unconsciously took priority for himself in the conception of new ideas. But after listening carefully, Wilbur accepted Orville's idea and then immediately improved upon it. Wilbur felt that the pilot already had a lot to do handling the other controls, so he suggested that the movable rudder be connected with the wing-warping mechanism so that the rudder would automatically deflect in the correct direction when the wings were warped. Within a few hours that morning, meshing their ideas, the Wrights developed an effective control mechanism for yaw on their flying machine—the deflection of the vertical rudder. With this, their flight control system was complete, and the 1902 glider became the first flying machine with full control around all three axes—pitch, yaw, and roll. After the movable rudder was installed, their tailspins and well digging disappeared.

The overall structural design of the glider—its lightness and durability—constituted the third technical contribution. They used lightweight, flexible, and resilient spruce wood that yielded but would not usually break when the glider had a hard encounter with the ground. The fabric of the wings, canard, and vertical tail played a structural role, being applied on the bias so as to add strength. Amazingly enough, they did not rigidly attach the wooden ribs and spars to each other; the cloth covering and pockets around the joints held the structure together—one reason why they could easily repair their machine whenever any damage occurred.

These contributions produced an excellent *system*. In the 1902 glider, the Wrights had good aerodynamics, good flight control, and good structures, and these elements all worked together to make the glider a tremendous success, the first successful *aeronautical system*. Their flying experiences in 1902 put Wilbur and Orville Wright far ahead of any other would-be inventor of the flying machine,

The Wrights' 1902 glider. Technically and in terms of number of flights, it proved to be a spectacular success. In comparison to their earlier gliders, the 1902 model soared almost effortlessly through the air. With its high-aspect-ratio wings, the machine appeared as an object of aesthetic beauty in flight. In modern aeronautical engineering, there is a well-worn saying that "an airplane that looks beautiful will fly beautifully." Although not always true, this description fit the Wrights' 1902 glider. Courtesy of the National Air and Space Museum.

although hardly anybody else knew it. But the Wrights knew it. Their technical maturity grew along with their self-confidence and determination to press ahead.

Unintentionally, the ten-day period following October 5 contrasted the Wrights' technology with that of Chanute. For the second year in a row, Chanute visited the Wrights' camp, this time bringing with him Augustus Herring and two flying machines. Herring, like Huffaker who had visited the Wrights' camp the year before, had worked for Samuel Langley at the Smithsonian, though only briefly. Since 1896 Chanute had employed Herring off and on to help build and fly gliders. Herring made a few marginally successful flights from sand dunes on the Indiana shore of Lake Michigan during the summers of 1896 and 1897. Now, in the summer of 1902, Chanute had again hired Herring to rebuild a multiple-wing glider of Chanute's design and to fly it at Kill Devil Hills, possibly with the help of the Wrights. Chanute also shipped to Kill Devil Hills a second flying machine, an oscillating wing machine that Charles Lamson built under contract.

Chanute and Herring arrived at Kill Devil Hills on October 5, and the Lamson oscillating wing machine showed up a few days later. This triplane had wings that could oscillate in the forward and backward directions, a feature that Chanute believed might enhance longitudinal stability. The multiple-wing glider, which had arrived two weeks earlier, was an ungainly looking machine with twelve wings having a total area of 150 square feet. On October 6 and 10, Herring attempted a few unsuccessful glides with the multiwing machine. Orville wrote in his diary on October 11: "Mr. Herring has decided that it is useless to make further experiments with the multi-wing. I think that a great deal of the trouble with it came from its structural weakness, as I noticed that in winds which were not even enough for support (winds with not enough velocity to lift the machine), the surfaces were badly distorted, twisting so that, while the wind at one end was on the underside, often at the other extreme it was on top." Orville went on to make a rather sad but important observation: "Mr. Chanute seems much disappointed in the way it works."

Octave Chanute, gracious old man of American aeronautics, dean and assimilator of the state of the art, and accomplished engineer, found his machines completely upstaged by the excellent performance of the Wrights' 1902 glider. Events on October 14 compounded the situation. "After breakfast we took the Lamson machine out in front of the building ready for gliding," Orville wrote, "but Herring soon decided to take it inside again to take its weight and ascertain its center of lift." The machine never flew. That same day Chanute and Herring rather abruptly left the camp to return home, leaving both machines behind. Embarrassed by the lack of performance of his machines in the face of the Wrights' obviously successful glider, Chanute later gave both his machines to the Wrights, a gift they graciously accepted but with which they never intended to do anything. When the Wrights left camp that year, Chanute's two machines remained behind, stored in the shed along with their 1902 glider. There they remained until they were destroyed in a violent storm late in 1907.

WORKING ON POWERED MACHINES

The Wrights left Kill Devil Hills on October 28. What a change from the year before! They now had the most successful flying machine ever developed, and they each had more than an hour of total flight time in the air. Their aeronautical technology had advanced far beyond that of any other—more than Langley, more than Chanute, even more than Lilienthal. Only one element was missing—propulsion. Until now, gravity had been their mode of propulsion—gliding downhill. In his 1899 letter to the Smithsonian, Wilbur had stated his original goal to "add my mite to help on the future worker who will attain final success." Now, he and his brother were about to become "the future worker," and they knew it. By the end of 1902, the Wrights had for all intents and purposes invented the airplane. All they needed for final success was an adequate propulsion device.

On December 11, Wilbur outlined their plans in a letter to Chanute: "It is our intention next year to build a machine much larger and about twice as heavy as our present machine. With it we will work out problems related to starting and handling heavy weight machines, and if we find it under satisfactory control in flight, we will proceed to mount a motor." Wilbur gave more details to George Spratt in a December 29 letter: "We are thinking of building a machine next year with 500 sq. ft. surface, about 40 ft. x 6 ft. 6 inches. This will give us opportunity to work out problems connected with the management of large machines both in the air and on the ground, such as starting, &c., &c. If all goes well the next step will be to apply a motor."

The end of 1902 also brought a change in the nature of the now voluminous correspondence between Wilbur Wright and Octave Chanute. Although such correspondence continued unabated, the roles became reversed. Previously, the Wrights learned from Chanute, particularly in regard to state-of-the-art aeronautics and how to make calculations for the design of a flying machine. Now, the Wrights had far surpassed Chanute in the fundamental understanding of heavier-than-air flight, and Chanute learned from the Wrights. It is hard to tell when and how much Chanute realized that this change had taken place. Much of his further correspondence to Wilbur continued to make technical suggestions that Wilbur knew were at best ill-advised and sometimes outright incorrect. In deference to their friendship, however, Wilbur patiently endured what became a growing feeling of exasperation with Chanute's apparent lack of ability to grasp some of the fundamentals that Wilbur shared with him.

The correspondence took another, more deliberate turn. After benefiting from the open aeronautic literature in their early work, the Wrights stifled the transfer of their technology to others. Recall that Wilbur was reluctant to have Chanute publish their tables of aerodynamic coefficients obtained from their wind tunnel tests. This attitude became more pronounced as the Wrights began to realize they actually were inventing the first successful airplane, and they became more conscious of protecting their invention. In the summer of 1903 Chanute prepared a paper on aeronautics for the French journal *Revue Generale des Sciences.* He wished to describe various aspects of the Wrights' 1902 glider and wrote several letters to Wilbur to verify certain technical features of the machine. Wilbur's replies were guarded. After reading a draft of Chanute's paper, Wilbur wrote to him on July 22, 1903: "The inaccuracies in the *Revue des Sciences* article in reference to our machine, to which I called your attention in a former letter, seemed to Orville and myself too serious to be allowed to appear in print." The next day, Chanute wrote asking questions about the operation of the vertical tail and sent the letter by special delivery. Wilbur quickly replied: "The vertical tail is operated by wires leading to the wires which connect the wing tips. Thus the movement of the wing tips operates the rudder. *This statement is not for publication,* but merely to correct the misapprehension in your own mind. As the laws of France & Germany provide that patents will be held invalid if the matter claimed has been publicly printed we prefer to exercise reasonable caution about the details of our machine until the question of patents is settled. I only see three methods of dealing with this matter: (1) Tell the truth. (2) Tell nothing specific. (3) Tell something not true. I really cannot advise either the first or the third course."

Chanute finally got the message. In a July 27 letter on, he wrote: "I was puzzled by the way you put things in your former letters. You were sarcastic and I did not catch the idea that you feared that the description might forestall a patent. Now that I know it, I take pleasure in suppressing the passage altogether. I believe however that it would have proved quite harmless as the construction (of a rudder) is ancient and well known." Chanute still did not grasp the significance of the coupled rudder. It became clear to both men, however, that the tone of their correspondence certainly was changing. We cannot condemn the Wrights for suddenly becoming more secretive. Patent rights are a serious matter, and with their realization that they were on the verge of inventing the first successful airplane, they can not be faulted for trying to protect the essential elements of their invention.

Before the end of 1902, the Wrights had already made some basic design calculations for their new, powered flying machine. When Wilbur noted in his December 29 letter to George Spratt that they were planning to construct a machine much larger than their 1902 glider, with a wing area of 500 square feet and a wingspan of 40 feet (compared to 305 square feet and 32 feet, respectively, for the 1902 glider), they had already calculated this wing size to account for the increased weight due to an engine. They estimated the engine weight to be about 180 pounds, thus giving them a total design weight of 625 pounds. Using the tables of lift coefficients obtained from their wind tunnel tests, they felt that a coefficient of about 0.8 could be achieved at a reasonable angle of attack. From the weight, wing area, and lift coefficient, the Wrights calculated that a flight speed of twenty-three miles per hour was necessary to lift off the ground. (See Appendix C for the details of this calculation.) The thrust from the engine-propeller combination would have to accelerate the machine to this velocity. Based on the wing drag coefficient measurements from their wind tunnel plus an estimate of the "framing resistance," that is, the drag of everything else—struts, wires, fuselage frame, pilot, and so on—the Wrights calculated that ninety pounds of thrust was required. Carefully and with a systematic methodology, the Wrights calculated virtually everything needed to design the new machine. They functioned as mature aeronautical engineers.

When Bishop Wright met Wilbur and Orville at the Dayton train station upon their return home on October 31, 1902, an air of success and optimism surrounded them, in contrast to the previous year. The brothers had already made the decision to build a new, powered machine. On December 3, a number of letters went out under the Wright Cycle Company letterhead to engine manufacturers, inquiring about the availability of a gasoline-fueled engine that could develop eight to nine horsepower and weighed no more than 180 pounds. All the replies were negative. The industry deemed such a high horsepower-to-weight ratio beyond the state of the art of engine manufacture at that time. Once again, the Wrights had to face doing it all by themselves, in the same spirit of their wind tunnel tests.

The brothers divided their responsibilities; Orville took charge of designing and building the engine, and Wilbur assumed the responsibility for the propeller design and construction. These self-imposed assignments reflected expedient and efficient engineering management—they considered it the best way to get the job done. Even with these focused assignments, the brothers continued to work as a team, sharing their thoughts and progress, each providing technical advice to the other.

Orville, along with Charlie Taylor, a mechanic they had hired in 1901 to relieve them of some of the bicycle shop work, began to design and build the engine in December. Taylor, the right man at the right time, played a role in the design of the engine and did virtually all the machining. No engineering drawings were made during its design; rather, they relied on sketches and Taylor's personal hand-crafting. They ran the first engine tests on February 12, 1903. A day later, Bishop Wright noted in his diary: "The boys broke their little gas motor in the afternoon." Dripping gasoline froze the bearings, breaking the engine body and frame. They received a new aluminum casting for the engine block on April 20 and success-fully ran the rebuilt engine in May. The Wrights felt optimistic about their engine. Wilbur wrote to George Spratt: "We recently built a four-cylinder gasoline engine with 4" piston and 4" stroke, to see how powerful it would be, and what it would weigh. At 670 revolutions per min. it developed 8 $\frac{1}{4}$ horsepower brake test. By speeding it up to 1,000 rev. we will easily get eleven horsepower and possibly a lit-tle more at still higher speed, though the increase is not in exact proportion to the increase in number of revolutions. The weight including the 30-pound flywheel is 140 lbs." According to their calculations of the power required for their new fly-ing machine, the performance of their handmade engine appeared marginal but sufficient.

A reciprocating engine itself does not produce forward thrust, which is needed to propel the airplane forward. Rather, the engine provides power in the form of a rotating shaft, and thrust is provided by a propeller connected to the engine shaft. Wilbur assumed the responsibility for the propeller design, although in the process he had many stimulating and sometimes heated intellectual discussions with Orville about the aerodynamic complexities of propeller design and per-formance. Between them, they produced the first viable propeller theory in the history of aeronautical engineering. Essentially a version of what today is called *blade-element theory*, it utilized aerodynamic information about airfoils from their previous wind tunnel tests. Wilbur recognized that a propeller is nothing more than a twisted wing oriented such that the "lift force" produced on this twisted wing is in the flight direction, that is, in the *thrust* direction. Orville summed up the results in a June 7, 1903, letter to George Spratt:

> During the time the engine was building we were engaged in some very heated discussions on the principles of screw propellers. We had been unable to find anything of value in any of the works to which we had access, so that we worked out a theory of our own on the subject, and soon discovered, as we usually do, that all the propellers built heretofore are *all wrong,* and then built a pair of propellers 8 $^1/_8$ ft. in diameter, based on our theory, which are *all right!* (till we have a chance to test them down at Kitty Hawk and find out differently). Isn't it astonishing that all these secrets have been preserved for so many years just so that we could discover them!! Well, our propellers are so different from any that have been used before that they will have to either be a good deal better, or a good deal worse.

For comparison, the crude propellers employed by would-be European flying machines inventors in the nineteenth century had efficiencies on the order of 40 to 50 percent. (Propeller efficiency is defined as the power output of the propeller divided by the shaft power input from the engine, expressed in terms of percentage.) Samuel Langley ran tests with his whirling arm on a propeller of his own design and measured a propeller efficiency of 52 percent; in *Experiments in Aerodynamics* Langley admitted that the "form of the propeller blades" was "not a very good one." In contrast, the Wrights predicted an efficiency of 66 percent for their propeller. Their efficiency actually turned out to be even better. An anonymous article entitled "Wrights' Propeller Efficiency," published in the November 1909 *Aeronautics* in New York, reported that a certain Captain Eberhardt in Berlin had taken detailed measurements of the propeller used by Wilbur in his European flights during 1908 and 1909. From this, Eberhardt measured a value of 76 percent for propeller efficiency. (In 2002, researchers at Old Dominion University, using a large wind tunnel at the NASA Langley Research Center in Hampton, Virginia, measured a phenomenal value of 84 percent for the Wrights' propeller efficiency.)

The Wrights made a quantum jump in the art of propeller design. Their blade-element propeller theory constituted a spectacular advancement over any existing means of designing propellers. Their fundamental understanding of the true aerodynamic function of a propeller formed the underlying basis of this theory, and their propeller theory exhibited a degree of technical maturity never seen before in applied aerodynamics. The details are laid out in the Wrights' notes as compiled by McFarland.[1] The importance of the Wrights' propeller theory in the history of early flight technology and the degree to which their highly efficient propeller design contributed to the success of the 1903 Wright Flyer are not always well recognized.

Following their now entrenched policy of doing everything themselves, the brothers carved two propellers out of laminated spruce, each 8.5 feet in diameter. They covered the tips with a thin layer of light duck canvas glued to the wood to keep it from splitting. Finally, they coated the propellers with aluminum paint. They were connected to the engine shaft via chain drives much like bicycle chains, arranged so that the propellers rotated in opposite directions to negate the gyroscopic effect. The whole propulsion system weighed about 200 pounds, only 20 pounds more than their original estimate.

The Wrights arrived at their Kill Devil Hills camp on September 25, 1903, and fourteen days later the crated parts of their new machine arrived. On October 9 they began to assemble it.

Meanwhile, 400 miles east of Dayton, at the Smithsonian Institution in Washington, D.C., Samuel Langley also feverishly worked on a large flying machine. After his stunning success with the steam-powered aerodromes in 1896, a hiatus occurred in his aeronautical research. After all, he had proven without a shadow of a doubt the technical feasibility of heavier-than-air powered flight—the fundamental goal of his earlier work. Not content to sit on this accomplishment, however, Langley wrote a letter to Octave Chanute in June 1897, stating: "If anyone were to put at my disposal the considerable amount—fifty thousand dollars or more—for . . . an aerodrome carrying a man or men, with a capacity for some hours of flight, I feel that I could build it and should enjoy the task." He went on to predict that he could accomplish this feat within two or three years from the time he would start. Behind the scenes in Washington, Langley became proactive in searching for a sponsor for such a large aerodrome, thinking that the U.S. government was his best chance.

Circumstances became more favorable with the approach of the Spanish-American War, and finally in 1898 Langley received an award of $50,000 from the War Department for the construction of a large, piloted flying machine.

Design of the machine started in earnest in May 1898 when Langley hired an assistant, Charles Manly, a young, new graduate from the Sibley School of Mechanical Engineering of Cornell University. Manly had great interest in the prospects for and the technology of powered flight. The two men got along very well. Manly soon became an instrumental force behind the design and construction of the new aerodrome.

When Langley hired Charles Manly, he tapped into the mainline profession of mechanical engineering as it existed at that time. Trying to find its clear identity, mechanical engineering endured being pulled in one direction by the older shop culture and in another direction by the school culture, which was beginning to produce college-educated engineers in growing numbers. Charles Manly graduated from one of the leading engineering colleges in the United States at that time, and he had as a professor and mentor the leading engineering educator Robert Thurston. Manly brought a certain cachet to Langley's aeronautical program, namely, that of the new breed of engineers steeped in a formal background of mathematics, physics, and analysis. Moreover, influenced by Thurston's interest in aeronautical technology, Manly focused on that aspect of flying machines.

The design philosophy behind the new aerodrome involved scaling it up from the previous successful aerodrome number five. This scaling became the Achilles heel of the machine; one cannot simply take a previous successful flying machine, make it four times larger, and expect it to perform as successfully. In Langley's time, scaling laws for both structures and aerodynamics did not exist.

Departing from his earlier use of steam, Langley correctly decided on a gasoline-fueled engine as the proper prime mover for an aircraft. By 1898 the internal combustion engine had proven itself superior to the steam engine for relatively lightweight power plants and experienced a rapid state of development in Europe and the United States. When Langley chose to use a gasoline-fueled internal combustion engine for his full-scale aerodrome, he could do so with some confidence. Improving rapidly, the technology appeared to be in hand. With Langley taking this direction, the Wright brothers clearly felt that they should do the same.

Langley estimated that he needed an engine that could produce about twelve horsepower and weigh no more than 120 pounds, and that he needed two of these engines to power the aerodrome. He first commissioned Stephan Balzer of New York to produce such an engine, but dissatisfied with the results, Langley eventually had Manly redesign the power plant. Langley encountered the same problem as the Wrights—the horsepower-to-weight ratio required for the engine extended beyond the standard state-of-the-art engine manufacture. Charles Manly, however, pulled off a technological miracle. His modifications resulted in a five-cylinder radial engine that produced 52.4 horsepower and yet weighed only 208 pounds, a spectacular achievement for that time. Using a smaller, 1.5-horsepower gasoline-fueled engine, Langley accomplished a successful flight with a quarter-scale model aircraft in June 1901, and he made an even more successful flight of the model powered by a 3.2 horsepower engine in August 1903.

Encouraged by this success, Langley stepped directly to the full-size airplane. He mounted this tandem-winged aircraft on a catapult in order to provide an assisted takeoff. In turn, he placed the airplane and catapult on top of a houseboat on the Potomac River. On October 7, 1903, with Manly at the controls, the airplane was ready for flight. The launching had wide advance publicity, with the press on hand to watch what might be the first successful powered flight in history. The next day the *Washington Post* reported:

> A few yards from the houseboat were the boats of the reporters, who for three months had been stationed at Widewater. The newspapermen waved their hands. Manly looked down and smiled. Then his face hardened as he braced himself for the flight, which might have in store for him fame or death. The propeller wheels, a foot from his head, whirred around him one thousand times to the minute. A man forward fired two skyrockets. There came an answering "toot, toot," from the tugs. A mechanic stooped, cut the cable holding the catapult; there was a roaring, grinding noise—and the Langley airship tumbled over the edge of the houseboat and disappeared in the river, sixteen feet below. It simply slid into the water like a handful of mortar.[2]

Manly escaped unhurt. Langley believed the airplane had been fouled by the launching mechanism, and he tried again on December 8, 1903. Again the aerodrome fell into the river, and again Manly was fished out, unhurt. The fouling of the catapult was blamed again, but some experts maintained that the tail boom cracked due to structural weakness. At any rate, that fiasco ended Langley's attempts. The War Department gave up, stating that "we are still far from the ultimate goal [of human flight]."[3] Members of Congress and the press leveled vicious and unjustified attacks on Langley. (Human flight was still looked upon with much derision by most people.) In the face of this ridicule, Langley retired from the aeronautical scene. He died on February 27, 1906, a man in despair.

Would Langley's aerodrome have had the capability of sustained, equilibrium flight had it been successfully launched? Possibly. On one hand, Langley made no experiments with piloted gliders to get the feel of the air. He ignored completely the important aspects of flight control. He attempted to launch Manly into the air on a powered machine even though Manly had no flight experience. On the other hand, after making ninety-three technical modifications to the Langley aerodrome, not the least of which included the addition of pontoons, Glenn Curtiss flew the aerodrome from Lake Keuka in upstate New York. However, with the modifications, the aerodrome was a different machine. Langley's aeronautical work nevertheless lent the power of his respected technical reputation to the cause of mechanical flight, and his aerodromes provided encouragement to others.

In terms of contributions to the technology of early flight, Langley's full-size aerodrome made none, except for its powerful, lightweight engine. Compared to the Wright brothers' technical maturity at that time, Langley's was almost retrograde. The aerodrome failed as a system; it had marginal aerodynamics, excellent propulsion, and marginal longitudinal and directional control (movable horizontal tail and vertical rudder controlled by the pilot.) It had no lateral control, and the structural aspect of the system proved a total failure. Even the excellent propulsion may have been a potential source of failure; had the aerodrome been able to stay together long enough for a flight, it would have been greatly *overpowered* and may not have been able to withstand the heavier airloads that would have resulted from flying too fast.

An interesting sidebar in the technology of early flight occurred in November 1901. Chanute corresponded with most of his contemporary aeronautical colleagues, including Langley. In November, Langley sent Chanute some data on a cambered airfoil, which Langley considered confidential; he asked Chanute not to publish the data. Chanute noted that the Langley airfoil appeared similar to some of the shapes that the Wrights were testing. He sent Wilbur Langley's letter, along with the shape

Samuel Langley's 1903 aerodrome. Langley designed and built this large piloted flying machine based on his successful 1896 aerodrome but made it four times larger. This photograph shows the aerodrome moments after launch on December 8, 1903; the rear wings have totally collapsed, and the aerodrome is in the process of flipping over on its back. Fortunately, pilot Charles Manly emerged unhurt from the wreckage. Langley's aeronautical work ended with this attempted flight. Nine days later, the Wright Flyer successfully flew at Kill Devil Hills. Author's collection.

of the airfoil and the corresponding aerodynamic data that Langley had measured on his whirling arm. (Recall that Langley and Huffaker had made measurements on cambered airfoils but never published the data.) Langley intended to use this airfoil for his aerodrome. Chanute asked Wilbur to examine the data, saying, "I shall be glad to know how your new experiments on surfaces agree with Langley's."

The Wrights went one better. They constructed a wind tunnel model identical to that tested by Langley, with an aspect ratio of four. They measured values of drift-to-lift ratios from angles of attack of zero to fifty degrees and compared them with Langley's data. The comparisons were for the most part within two percent—an incredibly good agreement for the technology of that day. The Wrights found, however, that the aerodynamic efficiency of Langley's airfoil compared poorly to that for some of their own airfoils. On November 14, 1901, Wilbur wrote to Chanute about the Langley airfoil: "This surface is by no means an efficient one for flying, as its best angle of gliding (a function of the lift-to-drag ratio) is 8.5°; and with framing and operator about 11°. Among the thirty or forty surfaces we have tested we found nearly all equal or superior to this one in lift and very much better in tangential (the axial force)."

How incredibly ironic! In late 1901 Wilbur and Orville actually measured the aerodynamic properties of an airfoil that was later used by Langley—their competition—on a flying machine. Moreover, they pointed out to Chanute the aerodynamic deficiencies of this airfoil. It is unclear whether or not Chanute ever transmitted this information to Langley. Chanute had violated Langley's confidence and, most likely, wanted to keep this quiet. Even if Chanute had transmitted the information, Langley would have ignored it. He was already in the final phases of the design of his large aerodrome and would have been reluctant to change the airfoil shape. Indeed, he wouldn't have known what change to make because the Wrights were certainly not going to give Langley the data for their airfoil shapes. Besides, Langley would not have accepted the Wrights' results on faith. Langley proceeded to use a relatively poor airfoil for his aerodrome, and the Wrights knew it!

How much contact and exchange of information took place between Langley and the Wrights during the design and construction of their respective flying machines? Not secretive about his work, Langley made known to many the progress on his aerodrome—in stark contrast with the Wrights' work on their powered 1903 flyer, known privately by only a handful of close friends, including Chanute and Spratt. Clearly during this time, the Wrights were well aware of Langley's work on his aerodrome. In turn, Langley slowly began to be aware of the Wrights' activities, mainly via Chanute. The Wrights' first contact with the Smithsonian—their letter of May 30, 1899, requesting basic information on aeronautics—was handled by Richard Rathburn, the assistant secretary. It is highly unlikely that Langley became aware of or paid any attention to this letter. Langley personally became aware of the Wright brothers for the first time in an exchange of letters between Wilbur, Chanute, and Langley in 1901. In a November 14 letter to Chanute, Wilbur thanked him for the gift of Lilienthal's book and then went on to ask: "Do you think Prof. Langley could be induced to reprint this book in the Smithsonian *Reports?* It is a wonderful book and should be given an English publication." Strangely enough, Chanute felt pessimistic about Langley's possible response. He replied to Wilbur: "I fear Langley will not favor republishing Lilienthal in English." Chanute neverthe-less transmitted this idea to Langley, who indeed declined. In Chanute's December 19 letter to Wilbur, he states, in a tone somewhat critical of Langley: "I enclose a let-ter from Prof. Langley giving his reason for not publishing Lilienthal in English. As it was my understanding that the professor does not read German, I did not think that the supervision of a translator would involve much labor for him. His own book has been practically ready for 3 or 4 years [a reference to Langley's writings that even-tually were published in his *Memoir*], and I sometimes have felt that he was keep-ing other students back by not printing it." On December 23, Wilbur replied to Chanute, expressing regret over Langley's attitude.

Relations between the Wrights and Langley got off to a bad start. It was dur-ing this time that the Wrights tested Langley's airfoil and found it lacking in aero-dynamic efficiency. In the fall of 1902, Chanute informed Langley about the Wrights' successful glider flights. This news aroused Langley's interest, made more acute by the impending trial of his own aerodrome. Langley wrote to Chanute on October 17, "I should like very much to get some description of the extraordinary results which you told me were recently obtained by the Wright brothers." Langley quickly followed with a second letter on October 21: "After seeing you, I almost decided to go, or send someone, to see the remarkable experiments that you told me of by the Wright brothers. I telegraphed and wrote to them at Kitty Hawk, but have no answer, and I suppose their experiments are over." Wilbur explained to

Chanute in a November 12 letter: "We received from Mr. Langley, a few days before we finished our experiments at Kitty Hawk, a telegram, and afterwards a letter, inquiring whether there would be time for him to reach us and witness some of our trials before we left. We replied that it could be scarcely possible as we were intending to break camp in a few days. He made no mention of his experiments on the Potomac." Langley became more anxious. Still communicating through Chanute and not directly with the Wrights, Langley wrote on December 7: "I should be very glad to hear more of what the Wright brothers have done, and especially of their means of control, which you think better than the Penaud. I should be very glad to have either of them visit Washington at my expense, to get some of their ideas on this subject, if they are willing to communicate them." Chanute communicated this invitation to the Wrights, with the comment that it "seems to me cheeky." The Wrights declined, on the excuse of pressing business.

Langley and the Wrights never met; the Wrights' lack of enthusiasm and dwindling respect for Langley was the main reason. Chanute never encouraged such a meeting, although he himself rather frequently visited the Wrights during this period. The Wrights' feelings for Langley were reflected in their lack of regret over the first failure of Langley's aerodrome on October 7, 1903. The following week, Wilbur wrote to Chanute: "I see that Langley has had his fling, and failed. It seems to be our turn to throw now, and I wonder what our luck will be."

The Wrights felt all along that the Langley aerodrome was greatly flawed, and his failures simply reinforced their own confidence in their aeronautical expertise. The general public, however, did feel discouraged by Langley's failures. Particularly after the second and final dramatic failure on December 8, 1903, the public image of flying machines reached a new low. The regents of the Smithsonian, so embarrassed by Langley's failures (they had never been supportive of the aerodrome work at the Smithsonian in the first place), forbade the secretary from doing any more aeronautical research. The government and the general public felt successful flight to be years away.

SUCCESS

Nine days after Langley's second failure, the Wright Flyer rose from the sands of Kill Devil Hills. Wilbur and Orville left Dayton on September 23, 1903, arriving at

Kill Devil Hills two days later. They began work on a new building to add to their earlier shed and also took the 1902 glider out of storage. On September 28, they executed between sixty and one hundred flights with their old glider, just to brush up on their piloting skills. On October 8 the parts for the new flying machine, which they had shipped from Dayton, arrived at the camp. Delayed and hampered by bad weather, including some strong storms, they took almost a month to assemble the Wright Flyer. During this time they continued to gain experience flying the 1902 glider. When they finished assembling the flyer on November 5, the weight of the machine had crept upward to a total of 750 pounds including the pilot. (Most airplanes designed during this century have ended up with final weights larger than the initial estimates—a fact of life for airplane designers due to the natural tendency to add extra features as the design progresses. This phenomenon goes all the way back to the Wright Flyer.) The total wing area measured 510 square feet with a wingspan of 40.33 feet, yielding a very reasonable wing aspect ratio of 6.4.

Unfortunately, there were more delays. While testing the engine on the same day that they finished the assembly of the flyer, the jerking of the unsteady rotation of the propellers damaged the propeller shafts. The shafts had to be returned to Dayton for repairs. Charlie Taylor tried to minimize the turn-around time, and on November 20 they were ready for another engine test. The engine still ran roughly, loosening the sprockets that held the shafts in place. Nothing seemed to keep the sprockets tight for any length of time. Finally, in frustration the Wrights glued the sprocket nuts in place, using bicycle tire cement. In a November 23 letter to Charlie Taylor, Orville noted: "Thanks to Arnstein's hard cement, which will fix anything from a stop watch to a thrashing machine, we stuck those sprockets so tight I doubt whether they will ever come loose again." This low-tech solution worked fine.

The Wrights had calculated that the drag on their machine at the takeoff speed of twenty-three miles per hour would be ninety pounds. Now, with the increase in weight, their calculations showed that a thrust of one hundred pounds would be required. The prospect that their engine-propeller system might not be able to produce this higher thrust worried the Wrights. So they felt elated when static tests of the propulsion mechanism yielded 132 pounds of thrust, far higher than what they had calculated as necessary. By November 23, the Wrights knew that success was imminent. Murphy's law prevailed, however; on November 28 one of the propeller shafts cracked during further testing of the engine. They decided not to waste any more time with the old shafts. Orville promptly left for Dayton to make new, larger, more durable shafts made from spring steel.

He arrived back at the camp on December 11, the new shafts in hand. By December 12, everything was ready. The Wrights had not had the chance to fly their new machine first as a glider, as they had originally planned. Such glider tests would have been in keeping with their cautious, engineering approach; they would have liked to check the flying qualities of this much larger and heavier machine and to get some practice at the controls before attempting a powered flight. But inclement weather and the bad fortune with the propeller shafts had delayed them throughout the fall, and now they were running out of time. They decided to fly the powered machine at the earliest possible moment, without the benefit of any glider trials. On December 14, the Wrights called witnesses from Kitty Hawk to the camp and then flipped a coin to see who would be the first pilot. Wilbur won. The Wright Flyer had no wheels, and for takeoff the Wrights had designed a detachable small wheeled dolly on which the flyer was mounted. The dolly rode on a sixty-foot launching rail laid on the sand; the rail was made from four fifteen-foot two-by-fours laid end to end, covered with a thin metal strip. As soon as the flyer lifted off the ground, the dolly would fall away.

Wilbur started the engine. In Orville's words: "I grabbed the upright the best I could and off we went. By the time we had reached the last quarter of the third rail (about 35 to 40 feet) the speed was so great I could stay with it no longer." Wilbur took off. The flyer suddenly went into a steep climb, lost speed, stalled, and thumped back to the sand. Wilbur admitted to "an error in judgment." He had deflected the elevator too much and brought the nose too high. This is perfectly understandable considering neither of the brothers had been able to obtain any flight time with the flyer as a glider and hence were not used to the controls of the bigger machine. The thumping back to the ground slightly damaged the front canard surfaces.

With minor repairs and with the weather again favorable, the Wright Flyer was ready on December 17. This time Orville had his turn at the controls. They again laid the launching rail on level sand. They adjusted a camera to take a picture of the machine as it reached the end of the rail. With the engine on full throttle, the holding rope was released, and the machine began to move. The rest is history, as portrayed in the opening paragraphs of this book. We have come full circle in our presentation of the technology of early flight.

One cannot read or write of this epoch-making event without experiencing some of the excitement of the time. Wilbur Wright was thirty-six years old; Orville was thirty-two. Between them, they had done what no one before had accomplished. By their persistent efforts, their detailed research, and their superb engineering, the Wrights had made the world's first successful piloted heavier-than-air flight, satisfying all the necessary criteria laid down by aviation historians. After Orville's first flight on December 17, they made three more during the morning, the last covering 852 feet and remaining in the air for 59 seconds. The world of flight—and along with it the world of successful aeronautical engineering—had been born!

The Wrights continued to fly in 1904 and 1905, each year with a better machine. They flew out of a cow pasture named Huffman Prairie, just outside of Dayton. They had a powered machine now and no longer had to rely on the winds of Kitty Hawk. (Huffman Prairie is now swallowed by Patterson Field, part of the massive Wright-Patterson Air Force Base.) Their 1905 machine performed so successfully that it could remain in the air until it ran out of fuel—a duration of approximately half an hour. Many aeronautical historians credit the 1905 machine as being the first *truly* practical airplane. Another three years would go by, however, before the world would appreciate what happened on that cold morning on December 17, 1903, at Kill Devil Hills. In August 1908, Wilbur made the first public demonstration of their invention (by now, an improved design labeled the Wright Type A); it took place at the Hunaudieres race track just outside of Le Mans, France. A month later Orville demonstrated the same type of airplane to the U.S. Army at Fort Myer in Virginia, just outside of Washington, D.C. Finally, the Wrights obtained full recognition for what they had done—they had invented the world's first practical airplane.

Epilogue

The invention of the airplane is one of the greatest feats of the twentieth century. Its roots, however, are primarily in the nineteenth century, when the concept of the modern configuration airplane first bloomed, and the technology of early flight became an identifiable and somewhat logical entity. The Wright brothers did *not* invent the airplane—they invented the *first practical airplane.* To borrow a well-worn phrase, to do so they stood on the shoulders of giants. During the course of this book, we have watched the technology of early flight develop from the misguided notions of the tower jumpers, the first concepts for flapping wing ornithopters, and theories that had no basis, to the more rational ideas of George Cayley and the parallel development of classical physics, especially the growing understanding of fluid dynamics and later the more specialized branch called aerodynamics. Progress toward the invention of a practical flying machine, however, suffered due to the lack of technology transfer between the academicians who developed the science of fluid dynamics and the craftsmen who were struggling to invent the flying machine. Indeed, as late as the beginning of the twentieth century, many viewed work on flying machines as the sphere of madmen. Nevertheless, men like Cayley, Alphonse Penaud, Otto Lilienthal, and Samuel Langley devoted themselves to the belief that heavier-than-air flight was possible. The product of their labors, although not a successful airplane, fostered the growth of what we now call the technology of early flight.

Why then were the Wright brothers successful when everybody before them failed? The answer is multifaceted. First, they took the best of the early technology of flight, threw away the rest, and had the technical maturity to know the difference. Then they applied a methodology to their engineering efforts—a methodology that they did not find in any books but rather that grew from their innate engineering talent. Their mechanical experience in building lightweight, efficient bicycles gave them an understanding of materials and construction that translated to their building flying machines. They carefully planned their course, adopting Lilienthal's "airman" philosophy that it is important to first get into the air on a machine and learn to fly long *before* putting an engine on it. The Wrights did very little ad hoc; instead, wherever they knew how, they calculated and estimated the performance of their system *before* designing and flying it. This ability to make reasonable calculations to help them intelligently design the system is a major reason why I feel that Wilbur and Orville were the first true aeronautical engineers. In the end, they really knew what they were doing!

In particular, their wind tunnel testing program in the fall of 1901 proved to be a pivotal aspect that led to a new, spectacularly successful glider in 1902, and in the next year to the powered Wright Flyer and ultimate success. Without these wind tunnel tests, the achievement of successful, powered human flight might not have happened as early as December 17, 1903, and it might not have happened to the Wright brothers. In this light, it is a good thing that the Wrights did not use the Lilienthal table correctly and that they used the wrong value of Smeaton's coefficient; that is, it is a good thing that their calculations disagreed with their measurements on their 1900 and 1901 gliders. Otherwise, the Wrights may not have been driven to discover "the right aerodynamics."

Whole books have been written to address why the Wright brothers were successful. One of the best is by Peter Jakab, to date the most in-depth study of the Wrights' process of invention.[1] But we can also look to others for some guidance. In a recent article, Tom Crouch succinctly states about the Wrights: "They were superb engineers, self-trained, who developed an extraordinarily successful research strategy that enabled them to overcome one set of challenging problems after another, the full extent of which few other experimenters had even suspected. They moved toward the development of a practical flying machine through an evolutionary chain of seven experimental aircraft: one kite [1899], three gliders

[1900, 1901, 1902], and three powered airplanes [1903, 1904, 1905]. Each of these aircraft was a distillation of the experience gained with its predecessors."[2] From a completely different quarter, the pioneer aircraft manufacturer Glenn L. Martin, in delivering the seventeenth Wright Brothers Lecture to the Institute of Aeronautical Sciences in Washington, D.C., on the occasion of the fiftieth anniversary in 1953 of the successful flight of the Wright Flyer, had this to say:

> The Wrights did not work in a scientific wilderness. They stood on the foundation erected by their predecessors in science. Yet all of these pearls did not make a necklace until the inspired jewelers came along.
>
> The Wrights brought to aviation the scientific approach to invention in which intuition and logic are used as a leavening for the diverse technical facts at hand. This method, of course, had prevailed for a century prior to the Wrights, but it was their course of action following a conclusion that made them immortals and their predecessors only dreamers and vain experimenters. Once having fashioned an idea, the Wrights proceeded to test it in scale form, either in their wind tunnel at Dayton or in a kite or glider at Kitty Hawk. This was their genius—no matter how sound or complete the idea, they were prepared from the start to modify it, to improve it, or to abandon it. For them there was no perfection.[3]

Had the Wright brothers never embarked on their aeronautical work, most likely the successful airplane would have been invented in Europe about ten years later. As it turned out, the first credible airplane flights in Europe took place in 1906 in an ungainly machine called the *14-bis,* developed and flown in Paris by the dashing Brazilian engineer, Alberto Santos-Dumont. The first flight, on September 13, covered only seven meters, but a second flight, on October 23, remained airborne over a distance of close to sixty meters and won the Archdeacon prize for the first flight of twenty-five meters. After this, Santos-Dumont fitted octagonal-shaped ailerons to the machine, and on November 12 executed a flight that covered 220 meters, staying in the air for 21.2 seconds. These constituted the first official powered airplane flights in Europe, and because the earlier success of the Wrights in 1903 had not been recognized by most people, Santos-Dumont was heralded by many as the first person to successfully fly a heavier-than-air, powered machine. For almost a year, the

French felt that the airplane had been invented in France. The *14-bis,* however, was in no way a practical aircraft. Moreover, by 1906 some of the Wright brothers' technology had spread to Europe via lectures given primarily by Octave Chanute while visiting the continent. It is likely that some of this information aided Santos-Dumont. Had the Wrights never existed, the airplane would have been invented by others sometime very early in the twentieth century.

Yet the Wright brothers *were* the first ones to do it, and they did it with dispatch, achieving their historic flight on December 17, 1903—just a little over four years after Wilbur wrote his fateful letter to the Smithsonian. The Wright brothers were the right people at the right time.

We mention a few interesting postscripts here. Sir Hiram Maxim in England lived to see the success of the airplane. Knighted in 1901, in the same year he became involved with the development of a lightweight gasoline engine. Along with A. P. Thurston, he designed during 1909–10 a new flying machine powered by a gasoline engine. A pusher biplane with steel tubing and Duralumin—a new aluminum alloy—for the fuselage framework, this airplane made the first use of Duralumin in aircraft structures. The airplane was of no consequence in light of the rapid developments after the Wright brothers' success. Sir Hiram Maxim worked almost to the end; he was granted two patents in 1916. On November 24, 1916, he died from bronchopneumonia at his home at Sandhurst Lodge in Streatham High Road, leaving an estate valued at 33,000 pounds.

After the success of the Wright brothers, particularly Wilbur's spectacular 1908 demonstration flights in Europe, and in the face of events building up to World War I, Otto Lilienthal enjoyed a posthumous skyrocketing of public acclaim in Germany. On June 17, 1914, the people of Berlin dedicated a large stone memorial with a winged figure at the top, commemorating Otto Lilienthal's pioneering contributions to flight. The fact that World War I loomed only two months off might have had something to do with the commemoration of someone now viewed as a national hero. Funded primarily by public subscription, the Germans asked Orville and Wilbur Wright for a contribution. Holding no value in memorials, the Wrights instead gave $100 to Otto's wife in 1912, who by that time was virtually impoverished. Other memorials followed. A circle of stones placed by the town's inhabitants in the 1930s ringed the fatal crash site on Gollengerg Hill. In 1954, the townspeople added a memorial stone. In 1932, as mentioned earlier, the Flying Hill in Lichterfelde became a major monument to Lilienthal. In 1940, the Nazi regime remodeled Lilienthal's grave site according to a design funded by the Lilienthal Association for Aeronautical Research. A memorial of a different sort was also commissioned under the Nazi government branch that strictly controlled literature; a third edition of *Bird Flight as the Basis of Aviation* was published in 1939, with a preface by the Gottingen University professor and researcher, Ludwig Prandtl. (Prandtl is held by many as the most famous aerodynamicist of the twentieth century.) Prandtl's preface provided a glowing appreciation of Lilienthal's work in aeronautics.

Returning to the old idea of human-powered flight, cyclist Bryan Allen flew successfully under his own muscle power on August 23, 1977. Pedaling the Gossamer Condor, a huge ultra-lightweight "cellophane" airplane, he won $50,000 for the first man-powered flight on a figure-eight course around two points at least a half mile apart. (The Gossamer Condor now hangs in the National Air and Space Museum of the Smithsonian Institution.)

The technology of flight grew exponentially during the twentieth century. Today, airplanes routinely fly faster than the speed of sound; a single aircraft can reliably and safely carry 400 or more people across the oceans. In their wildest dreams, Wilbur and Orville had no idea that their invention would evolve into such flying machines. And the march goes on. Nobody can predict with certainty what flight technology will be in the twenty-first century. We return to Glenn Martin, however, for some words of wisdom. In his seventeenth Wright Brothers Lecture, Martin stated with authority: "We have the aerodynamic knowledge, the structural materials, the power plants, and the manufacturing capacities to perform any conceivable miracle in aviation . . . But miracles must be planned, nurtured, and executed with intelligence and hard work. The miracles of the second half century will prove progressively more difficult, more costly, and more time-consuming. They will demand great skills for all of us."[4] Although made in 1953, when the future was the second half of the twentieth century, Martin's statement has lasting value. It not only applies to the "miracle" achieved by the Wrights at the beginning of the twentieth century but also to all aerospace engineers, managers, and enthusiasts who will carry aeronautics into the twenty-first century and beyond.

What have the Wright brothers wrought? We all wait anxiously to follow the continuously unfolding answer.

Appendix A: Langley's Power Law

Consider an airplane in steady (unaccelerated) level flight at a given velocity. A certain amount of power is required to overcome the drag and sustain the airplane in flight at the given velocity. The amount of power required is different for different velocities. The variation of power required with velocity is sketched generically in Figure A.1. This curve is a *power required curve* for a given airplane at a given altitude. Consider point *a* on the curve, associated with high velocity. Because the velocity is high, the airplane needs to be only at a small angle of attack to the flow in order to produce high enough lift to sustain it in level flight. The drag, at point *a,* however, is high because the velocity is high, and hence the power required to overcome this high drag is also large, as indicated by point *a.* As the velocity decreases, the drag also decreases, and the power required becomes smaller, as shown by moving to the left away from point *a* along the curve. At the same time, however, the angle of attack of the airplane must increase as the velocity decreases, in order to keep the lift equal to the weight of the airplane and sustain it in the air. Moving to the left along the curve in Figure A.1, the power required continues to decrease until the angle of attack becomes so large that the drag starts to increase in spite of the decreasing velocity. At this point, the power required curve goes through a minimum value and then increases as velocity continues to decrease. This increase in power required is reflected by point *b* on the curve. The portion of the curve around point *b* is called the "backside of the power curve." On the backside of the power curve, starting at point *b,* when the velocity increases, the power required actually decreases, as seen by moving to the right along the curve from point *b.*

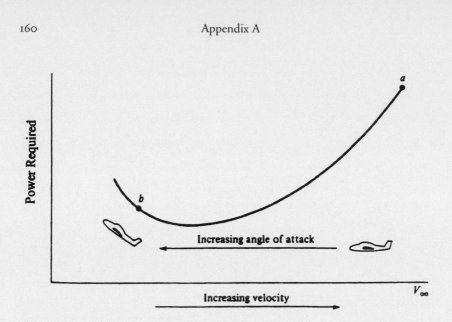

Fig. A.1. A generic power required curve for a given airplane in steady, level flight at a given altitude; a plot of power required as a function of flight velocity. Author's collection.

The power required curve for the flat plate models used by Samuel Langley, as calculated by the author, is shown in Figure A.2. Power required in watts is plotted versus airflow velocity in meters per second. The velocities for all of Langley's experiments ranged from eight to twenty meters per second. Note that this velocity range corresponds to the backside of the power curve in Figure A.2. Although Langley had no way of knowing, all his experimental data fell on the backside of the power curve, and his conclusion that it takes less power to fly faster was correct.

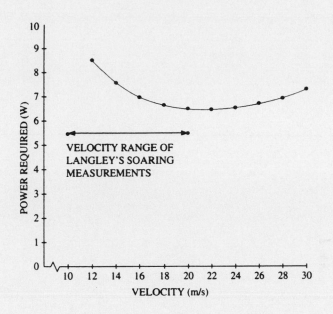

Fig A.2. The power required curve, calculated by the author, for Langley's flat plate models used in his whirling arm "soaring" experiments. Soaring meant that the lift generated on the model exactly balanced the weight of the model, achieved by adjusting the plate angle of attack for each different velocity. Note that all of Langley's measurements fall on the backside of the power curve, that is, that part of the curve where power required decreases with an increase in velocity. Author's collection.

Appendix B: How the Wrights Misinterpreted the Lilienthal Tables

The Wright brothers overpredicted lift for their 1900 and 1901 gliders because they misinterpreted and hence misused the Lilienthal tables. The main points are summarized in this appendix. The first misinterpretation has to do with Smeaton's coefficient. By the fall of 1901, the Wrights suspected the classic value of 0.005 for Smeaton's coefficient to be in error. In his paper to the Western Society, Wilbur states: "the well-known Smeaton coefficient of 0.005 V^2 for the wind pressure at 90 degrees is probably too great by at least 20 percent."[1] In 1891, Samuel Langley published his measurements for Smeaton's coefficient in his *Experiments in Aerodynamics,* giving an average value for his measurements of 0.003. Wilbur and Orville bought a copy of Langley's book in 1899 and became aware of Langley's measurements.

In both 1900 and 1901, however, while making the lift-and-drag calculations from the Lilienthal tables, they used the classic value of 0.005. They had the impression that Lilienthal used the value of 0.005 to obtain his force coefficients from his measured values of the actual force. To convert the tabulated coefficients back to force, the Wrights logically used the same value of 0.005, no matter what they might have thought the correct value of Smeaton's coefficient to be. In reality, Lilienthal did not use Smeaton's coefficient to reduce his data; instead, he formed his aerodynamic coefficients at different angles of attack by dividing the respective force measurements by his measured force on the wing at ninety degrees angle of attack. In so doing, the Smeaton coefficient, and any uncertainty thereof, simply divided out. The numbers in Lilienthal's tables are independent of any value of Smeaton's coefficient. The Wrights should have used 0.003 to obtain more accurate force data from the Lilienthal tables; instead, they used the higher value of 0.005.

Their second misinterpretation involved aspect ratio. Lilienthal made his aerodynamic force measurements on a model wing with an aspect ratio of 6.48. His table, therefore, can only be used to predict forces on wings with the same aspect ratio. (In 1918, Ludwig Prandtl in Germany published a theoretical method, the lifting line theory, for modifying aerodynamic data measured for wings of a given aspect ratio to apply to wings of another aspect ratio. At the time of the Wright's experiments, no such method existed.) The Wright brothers, however, readily used Lilienthal's table for their gliders, which had aspect ratios much smaller than 6.48. The Wrights acted as if they had no appreciation for the effect of aspect ratio on lift, even though Langley had published the first definitive experimental data on the effect of aspect ratio in his book *Experiments in Aerodynamics* and had conclusively shown that, at a given angle of attack, the lift coefficient decreases as the aspect ratio decreases. In 1900 and 1901, the Wrights appeared oblivious to this effect, which is not trivial.

Take the case of their 1900 glider, which had an aspect ratio of 3.5. Using Prandtl's lifting line theory, the lift coefficient for an aspect ratio 3.5 wing will be smaller than an aspect ratio 6.48 wing by the factor 0.814. The Wrights used an angle of attack of three degrees as their design point. They obtained a lift coefficient of 0.545 from Lilienthal's tables for a three degree angle of attack. They should have immediately multiplied this value by 0.814 to account for the aspect ratio effect. That is, for their aspect ratio of 3.5, the Wrights should have modified the value in Lilienthal's table to be (0.814)(0.545) = 0.44.

This number for the lift coefficient needs to be modified once more to account for the location along the airfoil of the maximum camber. Lilienthal's test wing had a thin airfoil in the shape of a circular arc, with maximum camber at the midchord location. The Wrights, on the other hand, placed their location of maximum camber near the leading edge because they felt this shape would reduce the movement of the center of pressure as the angle of attack changed. Because of the radically different maximum-camber locations, the Wrights' airfoils and Lilienthal's airfoil were aerodynamically different. The Wrights did not realize this difference; during their designs of the 1900 and 1901 gliders, they tended to view the numbers in the Lilienthal table as almost universal values for cambered airfoils, independent of wing shape and location of maximum camber. A more forward location of the maximum camber gives a smaller lift coefficient at a given angle of attack. At the design angle of attack of three degrees used by the Wrights, their forward camber location results in a smaller lift coefficient than for Lilienthal's circular arc at the same three degrees of attack. Calculating the change in the lift coefficient due to these different locations of maximum camber, we find that the already reduced lift coefficient of 0.44 discussed earlier should be further reduced by the amount of 0.11. Hence, at three degrees angle of attack, the Wrights should have used 0.44-0.11 = 0.33 as their lift coefficient.

Summarizing, for an angle of attack of three degrees, the Lilienthal table gives the lift coefficient as 0.545. The aspect ratio effect reduces that value to 0.44. The effect of location of the maximum camber further reduces the value to 0.33. The Wrights used 0.545 to calculate the lift for their 1900 and 1901 gliders; they should have used 0.33. On that basis alone, their calculations of lift were in error by a factor of 0.33/0.545 = 0.60. When the error in Smeaton's coefficient is factored into this calculation (the Wrights used 0.005, but should have used 0.003), their lift calculations were in error by a factor (0.003/0.005)(0.60) = 0.36. That is, the actual lift as predicted from the Lilienthal table, *with Lilienthal's value modified as discussed above,* gives a value one-third of that calculated by the Wrights, precisely what they observed experimentally, as recorded in Wilbur's diary on July 29, 1901: "Afternoon spent in kite tests. Found lift of machine much less than Lilienthal tables would indicate, reaching only about 1/3 as much."

This argument finally explains the apparent discrepancy between the Lilienthal table and the Wrights' measurements of lift on their glider; the numbers in the Lilienthal table and the Wrights' measurements were both reasonably valid. The Wrights' calculations, however, were not correct because they misinterpreted the numbers in the Lilienthal table and used these numbers incorrectly. In their defense, even if they had recognized the *qualitative* effects of different aspect ratios and different locations of maximum camber on lift coefficient, at that time they had no theoretical methods by which they could have *quantitatively* modified the Lilienthal data to apply to their gliders.

Appendix C: Wrights' Calculation of the Flyer Velocity

Using the equation

$$L = W = k\, V^2\, S\, C_L$$

where the lift, L, is equal to the weight, W, and k is Smeaton's coefficient, V is the flight velocity, S is the wing area, and C_L is the lift coefficient, the Wrights were able to estimate the velocity they would have to achieve to lift off the ground. Based on their previous flight experience and wind tunnel data, they felt that a reasonable lift coefficient at takeoff would be 0.8. They used the much more appropriate value of 0.003 for Smeaton's coefficient, which is for the case when V is in miles per hour. The result was

$$V = \sqrt{\frac{W}{k\, S\, C_L}} = \sqrt{\frac{625}{(0.003)(500)(0.8)}} = 23 \text{ mph}$$

Notes

Prologue: The Miracle of Flight

1. Orville Wright, "How We Made the First Flight," *Flying,* December 1913, p. 36.

2. A flying wing is a self-contained aircraft consisting only of a large wing. It has no identifiable fuselage and tail. The pilot, passengers, and baggage are located within the wing itself. To date, flying wings have been military aircraft; the F-117 stealth fighter and B-2 stealth bomber are flying wings.

3. Octave Chanute, *Progress in Flying Machines* (New York, 1894; reprint, Long Beach, CA: Lorenz and Herweg, 1976), p. 10.

4. Lord Kelvin's address to the British Association meeting, Oxford, August 1894, during a discussion of Maxim's flying machine.

5. These questions are discussed at length in John D. Anderson, Jr., *A History of Aerodynamics and Its Impact on Flying Machines* (New York: Cambridge University Press, 1997).

Chapter 1. Imaginings

1. Giovanni Alphonso Borelli, *De notu animalium* (Rome, 1680).

2. A thorough analysis of da Vinci's thoughts on aerodynamics is given in Anderson, *A History of Aerodynamics,* pp. 19–27.

3. Leonardo da Vinci, Codex Trivultianus, folio 6v, 1487–90.

4. Leonardo da Vinci, Codex E, folio 45v, 1513–15.

5. Leonardo da Vinci, *Codex Atlanticus.*

6. Isaac Newton, *Philosophiae naturalis principia mathematica,* 1687; English translation, *Mathematical Principles of Natural Philosophy,* trans. Florian Cajori (Berkeley: University of California Press, 1947).

7. Benjamin Robins, "Resistance of the Air and Experiments Relating to Air Resistance," *Philosophical Transactions of the Royal Society,* 1746.

8. Benjamin Robins, *New Principles of Gunnery Containing the Determination of the Force of Gunpowder and Investigation of the Difference in the Resisting Power of the Air to Swift and Slow Motions* (London, 1742); Robins, "Resistance of the Air."

9. John Smeaton, "An Experimental Inquiry Concerning the Natural Powers of Water and Wind to Turn Mills, and Other Machines, Depending on a Circular Motion," *Philosophical Transactions of the Royal Society of London,* vol. 51, 1759.

Chapter 2. Configurations

1. Jean Le Rond D'Alembert, l'Abbé Condorcet, and Charles Bossut, *Nouvelles Experiences sur la Resistance des Fluids* (Paris, 1777).

2. Cayley kept extensive notes in a series of notebooks now in the archives of the Royal Aeronautical Society, London. The aeronautical work of Cayley is nicely presented by C. H. Gibbs-Smith in *Sir George Cayley's Aeronautics, 1796–1855* (London: Her Majesty's Stationery Office, 1962).

3. Gibbs-Smith, *Sir George Cayley's Aeronautics,* p. 178. From a letter in the author's possession, written by Cayley's granddaughter.

4. Anonymous, from papers in the Library of the Royal Aeronautical Society, London.

Chapter 3. Experiments

1. Fred W. Breary, *5th Annual Report of the Aeronautical Society of Great Britain,* 1870, p. 11.

2. Francis H. Wenham, "Aerial Locomotion and the Laws by Which Heavy Bodies Impelled through Air Are Sustained," *1st Annual Report of the Aeronautical Society of Great Britain,* 1866, p. 19.

3. Duke of Argyll, *3rd Annual Report of the Aeronautical Society of Great Britain,* 1868, p. 9.

4. Mr. Moy, *3rd Annual Report of the Aeronautical Society of Great Britain,* 1868, p. 38.

5. M. De Lucy, "On the Flight of Birds, of Bats, and of Insects in Reference to the Subject of Aerial Locomotion," *4th Annual Report of the Aeronautical Society of Great Britain,* 1869, p. 84.

6. Anderson, *A History of Aerodynamics,* p. 130.

7. Hiram Maxim, *Artificial and Natural Flight* (London: Whittaker & Co., 1908), p. 31.

8. Ibid., pp. 3, ix.

9. Ibid., p. 8.

10. Ibid., p. vii.

Chapter 4. Aerodynamics

1. Anderson, *A History of Aerodynamics,* pp. 70-71.

2. C. H. Gibbs-Smith, *Sir George Cayley's Aeronautics,* p. 57.

3. Anderson, *A History of Aerodynamics,* p. 230.

4. Werner Schwipps, *Lilienthal: Die Biographie des ersten Fliegers* (Germany: Aviatic Verlag, 1979). English translation available in the library of the National Air and Space Museum (Washington, D.C.) in the form of a typed manuscript.

5. Ibid.

6. Ibid.

7. Ibid.

8. Ibid.

9. Octave Chanute, *Progress in Flying Machines,* p. 260.

10. S. P. Langley, *Experiments in Aerodynamics* (Washington, D.C.: Smithsonian Institution, 1891), p. 3.

11. S. P. Langley and C. M. Manly, *Langley Memoir on Mechanical Flight,* Smithsonian Contributions to Knowledge, vol. 27, no. 3 (Washington, D.C.: Smithsonian Institution, 1911), p. 2.

12. S. P. Langley, *Experiments in Aerodynamics,* p. 11.

13. Ibid., p. 9.

14. Ibid., p. 25.

15. Ibid., p. 3.

16. Anderson, *A History of Aerodynamics,* pp. 179–81, 456, 457.

17. Langley, p. 107.

18. Tom D. Crouch, *A Dream of Wings* (New York: Norton & Co., 1981), p. 55.

19. Langley and Manly, *Langley Memoir on Mechanical Flight,* p. 13.

20. Ibid., p. 20.

21. Ibid., p. 30.

22. Ibid., p. 85.

23. Ibid., p.108.

24. Crouch, *A Dream of Wings,* pp. 152–53.

Chapter 5. Technology: Collecting Data

1. Marvin W. McFarland, *The Papers of Orville and Wilbur Wright,* vols. 1 and 2 (New York: McGraw-Hill Book Co., 1953). Throughout chapters 5, 6, and 7, quoted material from the Wrights or from Wilbur's correspondence with Octave Chanute can be found in McFarland unless otherwise noted.

Chapter 6. Technology: Further Trials

1. Anderson, *A History of Aerodynamics,* pp. 209–16.

Chapter 7. Technology: A Working Aircraft

1. McFarland, *The Papers of Orville and Wilbur Wright,* pp. 594–640.
2. Quoted in Crouch, *A Dream of Wings,* p. 287.
3. Charles Gibbs-Smith, *Aviation: An Historical Survey* (London: Her Majesty's Stationery Office, 1970), p. 67.

Epilogue

1. Peter L. Jakab, *Visions of a Flying Machine* (Washington, D.C.: Smithsonian Institution Press, 1990).
2. Tom Crouch, "Kill Devil Hills, 17 December 1903," *Technology and Culture,* vol. 40, no. 3 (July 1999): pp. 594–98.
3. Glenn L. Martin, "The First Half-Century of Flight in America," *Journal of the Aeronautical Sciences,* vol. 21, no. 2 (February 1954): pp. 73–84.
4. Ibid.

Appendix B

1. Quotes from material written by the Wright brothers can be found in McFarland, *The Papers of Orville and Wilbur Wright.*

Suggested Further Reading

Anderson, John D., Jr. *A History of Aerodynamics.* New York: Cambridge University Press, 1997.

———. *Introduction to Flight,* 4th ed. Boston: McGraw-Hill, 2000.

———. *The Airplane: A History of Its Technology.* Reston, VA: American Institute of Aeronautics and Astronautics, 2003.

Combs, H. *Kill Devil Hill.* Boston: Houghton Mifflin, 1979.

Crouch, T. *A Dream of Wings.* New York: Norton, 1981.

———. *The Bishop's Boys.* New York: Norton, 1989.

Culick, Fred, and Spencer Dunmore. *On Great White Wings: The Wright Brothers and the Race for Flight.* New York: Hyperion, 2001.

Gibbs-Smith, C. H. *Aviation, An Historical Survey.* London: Her Majesty's Stationery Office, 1970.

———. *Sir George Cayley's Aeronautics, 1796–1855.* London: Her Majesty's Stationery Office, 1970.

Howard, Fred. *Wilbur and Orville: A Biography of the Wright Brothers.* New York: Alfred A. Knopf, 1987.

Jakab, P. L. *Visions of a Flying Machine.* Washington, D.C.: Smithsonian Institution Press, 1990.

McFarland, M. W., ed. *The Papers of Wilbur and Orville Wright.* New York: McGraw-Hill, 1953.

Index